Counseling Underachievers

Jeanne C. Bleuer

ERIC Counseling and Personnel Services Clearinghouse
2108 School of Education
The University of Michigan
Ann Arbor, MI 48109-1259

Second Edition, 1989

Third Printing, 1991

Cover design by Patricia Wisner

ERIC Counseling and Personnel Services Clearinghouse
2108 School of Education
The University of Michigan
Ann Arbor, MI 48109-1259

ISBN 1-56109-018-2

This publication was prepared with funding from the Office of Educational Research and Improvement, U.S. Department of Education under contract no. 400-86-0014. The opinions expressed in this report do not necessarily reflect the positions or policies of OERI or the Department of Education.

Educational Resources Information Center—ERIC

ERIC is a decentralized nationwide information system founded in 1966 and currently sponsored by the Office of Educational Research and Improvement within the U.S. Department of Education. It is the largest education related database in the world. ERIC is designed to collect educational documents and journal articles and to make them readily available through a number of products and services; e.g., the ERIC database, abstract journals, microfiche collections, online and CD-ROM computer searches, document reproductions, and information analysis publications. The ERIC audience is equally wide-ranging and includes teachers, counselors, administrators, supervisors, policy makers, librarians, media specialists, researchers, students, parents, and other educators and interested persons.

Counseling and Personnel Services Clearinghouse—CAPS

CAPS is one of the 16 subject-oriented clearinghouses of the ERIC system. CAPS' exceptionally broad coverage includes K-12 counseling and guidance, post-secondary and adult counseling services, and human resource development in business, industry and government. Among the topics addressed are:

- preparation, practice and supervision of counseling professionals
- development of theoretical constructs
- research on programs and practices
- interviewing and testing
- group work
- career planning and development
- employee assistance programs (EAPs)
- training and development
- marriage and family counseling
- student activities
- services to special populations (substance abusers, public offenders, students-at-risk)
- program evaluation

CAPS acquires literature in its subject area, processes the information into the ERIC database, and produces a variety of subject-specialized materials. It offers such products as monographs, special issues papers, state of the art studies, computer search analyses, bibliographies and digests. A quarterly newsletter (free upon request) features Clearinghouse activities, products, and articles on timely topics. CAPS professional staff also offers question-answering services, computer searching of the ERIC database, on-site user services with a complete ERIC microfiche collection at the CAPS Resources Center, and national, state and local workshops on high-priority counseling and human services concerns. We welcome visitors and mail or phone inquiries.

ERIC/CAPS
2108 School of Education
The University of Michigan
Ann Arbor, MI 48109-1259
(313) 764-9492

TABLE OF CONTENTS

INTRODUCTION

Ask school counselors to identify the most frustrating students to work with and underachievers probably will be near the top of the list. Motivating students to achieve is no easy task, and traditional counseling approaches are often ineffective in producing long-term behavior change. Underachievement is not simply a "they can, but they won't" situation, but a complex problem with both cognitive and affective dimensions produced by factors that are both internal and external to the student. For each individual, these factors interact in unique and complex ways, and a counseling strategy that works well with one student may not work at all with others. Because the problem is so complex, years of achievement motivation research have yet to produce a solution. Moreover, with the notorious gulf between research and practice, the practical implications of what researchers have learned have been virtually ignored. Consequently, the resources which counselors might use in working with underachievers are very few, and those that are available are scattered throughout the psychology and education literature.

Motivating students to achieve is no easy task, and traditional counseling approaches are often ineffective in producing long-term behavior change.

Counselors must also deal with the demands of parents and teachers, who often have high (and unrealistic) expectations regarding counselors' capacity to help students develop "achievement motivation." There may be a great deal of pressure to produce an accurate diagnosis and an effective treatment plan. Then, if the treatment plan doesn't work, parents and teachers question the counselor's competency.

Another difficulty is that, while parents and teachers tend to view the problem as residing primarily in the student, students often deny there is a problem and contend that whatever problem does exist belongs to their parents or teachers.

While the problem of underachievement has been around for a long time, the need for counselors to develop strategies for addressing the problem has never been more urgent. As schools heed the back-to-basics and educational reform movements and implement higher academic standards, students who met expectations in the past may not be able to continue to do so. Many parents are concerned about the impact this will have on their children's grade point averages and subsequent college and career options. Also, many educators (e.g., McDill, Natriello, & Pallas, 1985) are expressing concern that, without additional support services, the raised standards will alienate even further the ultimate underachievers—the potential dropouts.

Ironically, with the need for counseling services at an all-time high, many guidance and counseling programs are being severely cut back. As administrators attempt to juggle budgets in response to the new priorities, they see guidance and counseling as representing the "soft" side of the curriculum. Therefore, it is a prime target for reduced staffing and funding.

To address the problems raised by the educational reform movement, Walz (1984) advocates a proactive stance on the part of counselors. "Our current crisis in education, like all crises, confronts us not only with serious problems and challenges, but also with new opportunities" (p. 1). Counselors who see this crisis as an opportunity to define new counselor roles, acquire new competencies, and "bring their many talents and skills to the task can contribute enormously to a 'renaissance' in guidance and in the schools" (Walz, 1984, p. 2).

First among the new imperatives which Walz sees for school counselors is an increased emphasis on learning and recognition. Elaborating on this, Cole (1986) identifies "persistence in behaving like quasi-therapists" as an obstacle to excellence in counseling, because "time and again students and their parents tell us they want career and educational counseling" (p. 85). Neither Cole nor Walz is suggesting that counselors abandon their expertise in personal/social counseling. Many studies have shown that self-esteem and academic achievement are highly correlated, but there has been no definitive evidence to indicate which causes which. Therefore, a dual approach which aims at both building students' self-esteem *and* equipping them with effective learning strategies is the most effective method for helping underachievers improve their academic performance. Thus, what counselors need is to expand their expertise to include competence in helping students "learn how to learn."

The primary purpose of this guide is to present a first step in the direction of developing a comprehensive, systematic approach to counseling students for academic achievement. More specifically, it is intended to:

1. Emphasize the need for school counselors to place a high priority on and enhance their skill in academic achievement counseling.
2. Identify practical applications from research findings.
3. Present a preliminary model for counselor intervention.
4. Organize and present resources and ideas for implementing a comprehensive counseling intervention program.
5. Call attention to the need for further research and dissemination on exemplary programs for counseling underachievers.

Thus, what counselors need is to expand their expertise to include competence in helping students "learn how to learn."

In one sense, all of us are underachievers because "none of us uses more than 10% of our potentiality" (McHolland, 1971, p. 1). But, given 10% as the norm, we are not *expected* to do better. It is only when there is an "appreciable discrepancy" between performance and expected performance that underachievement is defined as a problem.

Schools, governments, and funding agencies have attempted to define underachievement in more specific terms in order to identify those students eligible for special services programs. For example, a New York state law defined the underachiever as "one who, on the basis of professional judgments, has not achieved for a year in accordance with his capacity" (*Guidance and the Underachiever*, 1967, p. 1). Other definitions have focused not so much on the time factor, but on the specific amount of discrepancy as determined by grade point averages and scores on standardized aptitude tests. "In the research literature, underachievement is usually defined in terms of a discrepancy between observed and expected academic performance. Often, grade-point average (GPA) is taken as a measure of performance and compared to the GPA predicted on the basis of measured mental ability. When the actual GPA falls about one sigma below the predicted GPA the student is labeled as an 'underachiever'" (Bednar & Weinberg, 1970, p. 1).

. . . underachievement is usually defined in terms of a discrepancy between observed and expected academic performance.

Obviously, the "labeling" of underachievers can be inaccurate, whether it is based on subjective judgment or objective criteria. The variability of individual teacher practices and school policies regarding grading, the validity and reliability of aptitude/ability tests in terms of socio-cultural biases, and the cumulative effect of underachievement on later aptitude/ability test performance are all factors that can lead to erroneous conclusions about students' actual aptitude and achievement. Fortunately, there has been little use of such specific criteria for the past several years and the identification of underachievers has been based more on parent and teacher observation than on the GPA/test score comparison.

In identifying signs of underachievement, McGuire and Lyons (1985) advise that "underachievement is frequently associated with some amount of behavioral disruptiveness as the underachiever is often frustrated or bored and thus more easily distracted than his or her achieving peers" (p. 37). On the other hand, Caliste (1980) cautions, "the child who makes himself inconspicuous in class, while not disturbing others, may be missing as much as the child who annoys peers or who requires the use of external control" (p. 3).

Whatever criteria are used to identify underachievers, a problem exists for school counselors if a parent or teacher *thinks* that a student's underachievement is a problem. In other words, the parent of a gifted student who drops to a B in one class can (and often does) demand as much attention from the counselor as the parent of an average student who is in danger of failing a course.

HOW SERIOUS IS THE PROBLEM?

Compared to many of the problems which today's school counselors must deal with, such as youth substance abuse, teen suicide and depression, teen pregnancy, family violence, and students at risk of dropping out of school, the problem of student underachievement may seem relatively minor. In fact, until fairly recently, the concept of underachievement was applied primarily to the gifted underachiever, and the costs of underachievement were viewed in terms of students wasting their talents and society being deprived of the potential contributions of these students' leadership and expertise. Today, however, there is a growing realization among counselors and educators that underachievement can be a serious problem for students at all ability levels and that the consequences of underachievement can be quite costly, both to the student and to society in general.

Today, however, there is a growing realization among counselors and educators that underachievement can be a serious problem for students at all ability levels. . . .

Underachievement often results in an alienation of students from the important adults in their lives. It can also lead to lowered student self-esteem because students who fall behind in class work do, in fact, miss out on important learning. Eventually they are unable to complete required assignments, even when they try. For these students, school then becomes a very unpleasant place to be because it reinforces daily their inadequacies and failures.

Alienation from adults, combined with low self-esteem, makes students highly susceptible to involvement in the more serious types of behaviors described above. This is particularly relevant to the current dropout problem. According to nearly all the recent studies, most dropouts are of average intellect, and "the stereotype of the dropout as an incapable, low-IQ juvenile is simply inaccurate" (Beck & Muia, 1980). In one study, it was found that over 65% of the dropouts were in the bottom quartile of their class according to their academic rank, but only 44% were in the bottom quartile of their class according to scholastic ability as measured by standardized tests (*The West Virginia Dropout Study*, 1987). In study after study, falling

. . . "the stereotype of the dropout as an incapable, low-IQ juvenile is simply inaccurate."

behind in school work has emerged as a major factor leading to early school leaving.

Students' Cost/Benefit Analyses

It is quite common for parents, teachers, and counselors to identify lack of motivation as the major cause of underachievement. However, as Whitmore (1986) points out, "it is never accurate to say that a student is unmotivated; the fact is, the child may be unmotivated to do specific schoolwork assigned but is motivated to engage in other learning activities or more rewarding alternatives, such as social interaction or daydreaming" (p. 67).

A review of the ERIC literature on underachievement, particularly teachers' and counselors' descriptions of underachievers and their behavior, suggests that many underachievers conduct a kind of cost/benefit analysis and choose the option that seems most attractive to them at the time. The costs of achieving may include expended effort, possible frustration, time away from other activities (friends, hobbies, entertainment), and potential alienation from friends who are non-achievers. The benefits of achieving may include higher grade point average, impressing friends, teacher and parent approval, increased educational opportunities, expanded career development options, and self-satisfaction. On the other side, the benefits of not achieving would be the positive aspects of the achievement costs (e.g., relaxing, being with friends), while the costs of not achieving would be the negative aspects of the achievement benefits (e.g., lower grade point average, parent disapproval). As McHolland (1971) points out, many underachievers are products of the "instant, turn-on, television-oriented culture" who want their rewards *now*. In this context, the underachiever's perception of the costs and benefits might look like this:

. . . many underachievers are products of the "instant, turn-on, television-oriented culture" who want their rewards now.

	Cost	Benefit
Achievement	Immediate	Delayed
Non-Achievement	Delayed	Immediate

Thus, students who have poor impulse control or place minimal importance on the longer range benefits of achievement are likely to opt consistently for non-achievement behavior (immediate benefit/delayed cost). This suggests that effective counseling interventions with these types of underachievers would include strategies to: (1) increase or enhance the immediate rewards of achievement; (2) help the student place greater value on the longer range rewards of achievement; and/or (3) increase the costs and immediacy of non-achievement. For maximum effectiveness, a comprehensive approach that incorporates all three strategies would be most desirable.

Effort vs. Ability

Another common assumption regarding underachievers is that they have the ability, but they simply won't put forth the effort. By definition, this is true, but perhaps the definition of ability shouldn't be limited to intelligence and aptitude. Sometimes, underachievement is the result not of lack of effort, but of ineffective learning strategies and study skills. When this is the case, criticism from parents and teachers can be quite devastating and lead to further underachievement because students eventually assume they *don't* have the ability to achieve, so why try. This is a good illustration of the need for counselors to identify the specific reasons for each student's underachievement and to develop counseling strategies accordingly.

Sometimes, underachievement is the result not of lack of effort, but of ineffective learning strategies and study skills.

Internal vs. External Factors

The key dimension of the effort vs. ability issue is that of control. It is usually assumed that underachievers have power in that they maintain control over whether they will engage in achievement behaviors. In most cases, this is probably true. Sometimes, however, the underachiever is more of a victim than an agent, particularly, for example, when strong external factors (poor home environment, negative peer pressure, uncomfortable school environment, etc.) work against the student's efforts to achieve. This can also be the case when unresolved psychological (i.e., internal) factors are influencing the student's behavior.

Psychological Factors. Some underachievers use school failure to avoid the anxiety associated with progressing to the next developmental stage of increased independence. Roth & Meyersburg (1963) have labeled this the "non-achievement syndrome" and describe these students as having a preadolescent-latency fixation. In contrast to this type of student is the adolescent who *is* trying to assert his/her independence. Goodstein (1980) identifies this as the "adolescent reaction" type of underachievement in which students "seek to define themselves in ways that often upset the parents. What previously may have been a 'good' boy or girl is now a person who is often intolerable. Rebellion often becomes the order of the day" (p. 107).

Whitmore (1986) identifies two additional psychological factors relevant to underachievement—fear of failure and fear of success. If a student doubts his/her ability, not putting forth the effort to achieve is one way to avoid putting that ability to the test. On the other hand, a student may fear success because it will likely be followed by increasingly higher expectations in the future. When underachievement appears to be a function of developmental or psychological factors such as these, individual counseling or group therapy may be necessary to effect change.

Family Factors. Zuccone and Amerikaner (1986) point out that "specific problems of individuals, especially children, are symptomatic of and often symbolic of disturbances in overall family functioning . . . the child is both an individual and part of a larger family system within which his or her poor school performance may be understandable or even purposeful" (p. 590).

Aside from dysfunctional family systems, even normal families may fail to provide children with adequate role modeling and support. McGuire and Lyons (1985) report that, in their experience, "frequently households of referred underachievers have no quiet place or time set aside for study and there was seldom any consistent attempt to monitor these children's homework. In a number of cases, the child's obligations in the family preempted time or energy from homework. . . . Moreover, educators and counselors often conclude that the families of underachievers are also unmotivated as they are often reluctant to attend school meetings or work collaboratively with teachers and other professionals" (p 38).

In an extensive review of the literature on parent involvement in education, Sattes (1985) concludes: "The literature on school-family relations is consistent. Meaningful parent involvement results in gains in student achievement and such related factors as attendance, motivation, self-concept, and school behavior. We have examined parent attitudes and home environment variables that are correlated with achievement. Reading to children, having books in the home, positive parent attitudes toward school, and high parental expectations for achievement are positively related to school achievement. . . . The evidence presents a powerful argument for home-based or parent-involved preschool programs for all children, because never again in a child's school career can a program result in such permanent and significant positive effects" (p. 22).

For school counselors, working with individual families may not be an option—in terms of time, family counseling skills, and/or school policy. Conducting parent education groups is one strategy that many counselors are using to help parents become more involved in their children's education. However, the parents who tend to participate in these programs are usually the ones who are already doing everything they can to help their children do well in school. When the home environment is obviously contributing to students' lack of achievement and working with the parents is apparently impossible, the best approach is to involve the students in group counseling. In this setting they can provide support for each other and can share strategies for dealing with their family problems.

Peer Influences. As all school counselors know, a student's "worst enemies" in terms of overcoming underachievement behavior may be his/her own peer group. Peers who are poor students themselves may actively discourage their friends from doing well, and peers who are good students may be insensitive to the underachiever's need to spend more time studying. A romantic relationship can leave any student with little time to study and little capacity to concentrate on school work. Whitmore (1986) points out that gifted students may underachieve intentionally to avoid the social penalties of peer alienation.

. . . gifted students may underachieve intentionally to avoid the social penalties of peer alienation.

On the other hand, peers can also be a powerful positive force. That is why group counseling has been shown to be much more effective than individual counseling in treating problems of underachievement. Even in working with individuals, counselors might find it useful to involve one or two of the underachiever's friends in setting up mutual goals, monitoring each other's progress, and celebrating each other's achievements.

School Factors. There is very little research that directly addresses the extent to which schools contribute to students' underachievement problems. However, the surge of dropout studies and educators' growing concern with identifying and meeting the needs of students at risk may reveal some important information. For example, in one study (Thomas, 1985) some students said they dropped out because: "No one cared why I was absent" or "Tried to see a counselor—they were always too busy" (p. 13).

Size of school, length of school day, attendance and discipline policies, scheduling flexibility to accommodate individual student needs, teachers' commitment to students, teacher-student interaction, extra-curricular activities, career guidance resources, and counselor-student ratio are among the many factors that can affect student achievement. Perhaps in the absence of definitive research in this area, it would be worthwhile for counselors to conduct surveys of their own students regarding their perceptions of these variables. The instrument could be a simple checklist with a like-dislike scale and room for comments and suggestions. The information collected could help counselors better understand students' individual and collective perceptions of their school environment, and also help administrators evaluate the strengths and weaknesses of the school.

Review of Achievement Motivation Theories

In the past twenty years, an extensive amount of research has been conducted on achievement motivation, particularly as it pertains to academic achievement. However, there appear to have been only a few attempts to examine this research for its relevance to counseling underachievers. In this section, a brief description of five current achievement motivation theories will be presented followed by a discussion of how these theories and the studies based on them can contribute to our understanding of why students underachieve.

Expectancy-Value Theory. According to Atkinson's (1964) theory of achievement motivation, individuals are motivated by two factors—the expectancy of attaining the goal, and the value they place on the goal object. If these two factors are both positive, individuals "are more likely to voluntarily approach achievement situations, perform tasks with greater intensity, persist in the face of failure, and select tasks of intermediate difficulty" (Stipek & Weisz, 1981, p. 117). Studies based on this theory have shown that individuals with high achievement motivation are motivated more by the hope of success than the fear of failure, while individuals with low achievement motivation are motivated more by fear of failure.

Locus of Control. This well-known and widely-researched concept described by Rotter (1966) focuses on the extent to which individuals view events as either contingent upon or independent of their own behaviors. Individuals who develop a "generalized expectancy" that the locus of control (LOC) is contingent upon their own behavior (i.e., ability and effort) have an *internal* locus of control. Those who believe that outcomes or events are caused by factors beyond their control (luck, chance, task difficulty, powerful others) have an *external* locus of control. Some studies classify people as either internals or externals, while others use instruments that treat the construct as a continuum. In terms of predicting achievement behavior, Rotter (1975), like Atkinson (1964), emphasizes the importance of the individual's perception of the *value* of the expected outcome in addition to his/her perceived locus of control for it. For example, a student who knows that he/she can get a high grade by studying still may not study if the high grade is not valued.

. . . a student who knows that he/she can get a high grade by studying still may not study if the high grade is not valued.

A positive correlation between academic achievement and internal locus of control has been well documented in the locus of control research. Therefore, since internality is the favored state, the literature often refers to subjects as I+ (internal LOC) and I- (external LOC) as used in Crandall's Intellectual Achievement Responsibility (IAR) Questionnaire (Crandall, Katkovsky, & Crandall, 1965). It would be useful to educators and counselors to know if internal LOC causes high academic achievement or vice versa. However, most studies have been cross-sectional and correlational, thus not able to determine causality. The few causal analyses that have been done have been contradictory. For example, Calsyn's (1973) and Stipek's (1980) cross-lagged panel correlation analyses both suggest that an internal locus of control causes achievement. However, in a quasi-longitudinal study, Kifer (1975) found that, even though both successful and unsuccessful students started out with internality in grade 2, internality increased through grades 4, 6, and 8 for successful students, but not for unsuccessful students. This suggests that higher achievement leads to higher internality.

Kifer also found that self-concept started out high for both groups in grade 2, but by grade 4 had already dropped significantly for unsuccessful students. Because of the conflicting evidence regarding causality, Marsh (1984) advocates a "dynamic equilibrium model" in which "academic achievement, self-concept, and self-attributions are interwoven in a network of reciprocal relations such that a change in any one will produce other changes in order to reestablish an equilibrium" (p. 1307).

Attribution Theory. In locus of control theory, both ability and effort are categorized as internal variables. However, in academic achievement research, it is important to distinguish between attributions made to effort and those made to ability and whether these are different for success and failure situations. Weiner's (1979) attribution theory builds on locus of control theory to identify three dimensions of causality: locus, stability, and control. Locus refers to internal vs. external causality. Control is separated from locus and distinguishes between controllable and uncontrollable factors. Thus, effort is an internal, uncontrollable factor. Stability classifies factors as either stable (fixed) or unstable (variable). Ability and task difficulty are examples of stable factors, and effort and luck are examples of unstable factors. The original attribution model introduced by Weiner, et al. (1971), and the basis for most attribution theory research, was a 2x2 matrix consisting of the four attributions: ability, effort, task difficulty, and luck. In essence, the control dimension isolates effort from all other factors because external factors are generally beyond the individual's control.

	Internal	External
Stable	Ability	Task Difficulty
Unstable	Effort	Luck

According to this theory, achievement behavior is determined largely by the causes to which individuals attribute their previous successes and failures. For example, if a student attributes previous academic success to high effort, he/she will tend to persist at new tasks, even in the face of failure. However, if a student has experienced continual failure in spite of persistent effort, he/she will eventually attribute those failures to low ability and will avoid new tasks or will put forth minimal effort before giving up. This condition has been described as "learned helplessness" (Dweck & Reppucci, 1973; Maier & Seligman, 1976). Dweck (1975) has demonstrated that "attribution retraining" (AR) can be used to alter the way that learned helpless children respond to their failures. In AR, training materials are arranged so that the child's increased efforts result in improved task performance. Then, periodically, the child is exposed to "failure" (an insoluble problem or too much to accomplish in the allotted time). The child is taught to attribute these failures to insufficient effort. Licht (1983) suggests an expansion of the AR model by focusing on the attribution of failure to ineffective task strategies. Although her suggestion was aimed at helping learning disabled children, the concept seems equally useful for underachievers in general who may lack or fail to use effective study skills.

In attribution studies, "actors" are subjects who are asked to make attributions regarding the causes of their own successes and failures, and "observers" are subjects who are asked to attribute causes for the successes and failures of others. Ross (1977) has described "the fundamental attribution error" as the tendency of observers to identify internal causes (i.e., to blame actors) for failures. Actors, on the other hand, tend to identify external causes (i.e., to make excuses) for their own failures. Extending this concept, Pettigrew (1979) has described "the ultimate attribution error" as the tendency of observers to blame dissimilar others, but excuse similar others for failure.

Self-Efficacy. Bandura's (1977) self-efficacy theory is grounded in locus of control and attribution theory, but places more emphasis on the importance of the distinction between contingency and control. According to this theory, the individual's "'efficacy expectations'—his confidence that he can successfully execute the behavior to produce certain outcomes" (Bandura, Adams, & Beyer, 1977, p. 126) will have a greater impact on achievement behavior than the outcome expectancies *per se* (i.e., the contingencies).

In his studies of the effects of attributional feedback to children, Schunk (1983) found that positive feedback to children about their ability led to higher achievement and self-efficacy than did no feedback, positive feedback about their effort, or even positive feedback about both their effort and their ability. This is consistent with research studies (e.g., Kun, 1977) that provide evidence that children use inverse compensation to judge effort and ability—i.e., the greater the effort, e.g., "You really worked hard," is interpreted as, "You had to work hard because you don't have high ability."

Personal Causation. According to deCharm (1968), individuals see themselves as either "origins" or "pawns." An origin is a person who feels that "what he is doing is the result of his own free choice; he is doing it because he wants to do it" (deCharm, 1968, p. 381). A pawn sees himself/herself as an instrument of others. This theory differs from LOC and attribution theory in that the focus is on the individual's perception of who controls his/her behavior, while the other theories focus on who or what controls outcomes or events.

Implications for Counseling Underachievers. Although achievement motivation theories may appear to be statements of the obvious, they help us to think more carefully about what we otherwise take for granted. For example, there is common consensus that "perceived ability" is a major prerequisite for motivation. The key word here is "perceived." Too often, adults assume that just because students have ability they will know they have it. For many underachievers, this may not be true, particularly if they have attributed past successes to external factors, such as easy courses, help from parents or teachers, etc. Exploring students' perception of their abilities (through measurement and/or discussion techniques) and helping them arrive at an accurate self-assessment can be an effective first step in counseling underachievers.

The achievement motivation theories and research also call attention to the fact that students' perceptions are based on their own past experiences. In other words, just *telling* students that they have the ability to achieve has little impact. They must, themselves, be helped to attribute their past successes to internal (ability plus effort) causes.

In addition to feeling that they *can* achieve, students must truly *want* to achieve. Therefore, their perceived value of the goal object (good grades, college admission, etc.) is a critical component of their motivation to achieve. Here again, this value must be "owned" by the *student*, not just by his/her parents and teachers. Immediate rewards, such as earning money for high grades, may work temporarily and they may help students experience success. These experiences can be useful if they are used to demonstrate to students that they *can* achieve. In the long run, however, values clarification, career exploration, and career planning activities are likely to be more effective because they can help students develop more conscious linkages between academic achievement and their own life/career goals.

In regard to the debate over whether internal locus of control causes high academic achievement or vice versa, basic counseling philosophy (i.e., belief in the individual's ability to change behavior) would probably support the position that locus of control is a situational, thus modifiable variable rather than a fixed personality trait. If this is so, early intervention to assure academic success from first grade on is extremely important. For counselors, the "dynamic equilibrium model" (Marsh, 1984) provides a very useful approach to conceptualizing the inter-relationships of locus of control, self-concept, and academic achievement. Using this model, counselors can aim for maximum intervention

Exploring students' perception of their abilities . . . and helping them arrive at an accurate self-assessment can be an effective first step in counseling underachievers.

In addition to feeling that they can achieve, students must truly want to achieve.

. . . counselors can aim for maximum intervention effectiveness by simultaneously implementing strategies designed to improve learning skill, enhance self-esteem, and increase internality.

effectiveness by simultaneously implementing strategies designed to improve learning skill, enhance self-esteem, and increase internality. Further research on this model could provide strong support for developmental guidance programs and primary prevention techniques.

In investigating the implications of attribution theory for peer counseling approaches, Rittenhouse, Stephan, and LeVine (1984) found that the peer counselors in their study tended to make the fundamental attribution error in giving advice to their peers. That is, they tended to "blame" the underachievers for their lack of achievement and gave advice that focused primarily on internal changes the students should make. This tendency very likely applies to parents and teachers of underachievers as well. In other words, as "observers," parents and teachers are likely to see the problem residing entirely in the child, whereas the child (the "actor") may be more aware of the external causes for his/her underachievement—some of which may be quite valid, such as a disturbing home environment or the need to work part-time.

One of the most useful contributions of attribution theory to counseling underachievers is the emphasis on the importance of helping students attribute both their successes and failures to internal causes, with lack of effort being the primary cause of failure. This concept is particularly important in providing tutoring and support services for underachievers. A tutor who spoonfeeds a student may help him/her through an immediate crisis, but in the long run does the student a disservice. Students who receive this kind of support are likely to attribute any successes to the tutor (an external cause) rather than to internalize the success.

"Perceived control" is an important concept that cuts across all of the theories. What counselors and teachers must remember, however, is that the potential for control must be both realistic and personally experienced by the student. Trying to convince a student "you can do it," when insurmountable barriers remain (e.g., the student is already too far behind, the home situation is too distracting) is unfair and may be quite destructive. Therefore, it is critical that counselors first explore the potential barriers to each student's underachievement and develop an intervention plan that will comprehensively address all of the relevant causes. Also, simply telling students that they are "in charge" will have very little impact compared to leading them through an analysis of the causes and effects of past behaviors (both successes and failures) and structuring experiences that demonstrate to them that, in fact, they *can* be in charge of their own achievements. Here, deCharm's (1968) origin/pawn concept can provide a useful topic for discussion. For example, with the adolescent reaction type of underachievers, counselors can urge students to analyze whether, in their acts of rebellion, they are really doing what *they* want to do or are caught up in psychological struggles with others in which their reactions are making them behave more like pawns than origins.

Implicit throughout all of the achievement theories is the concept of personal responsibility for outcomes —both immediate and long-range. This concept is quite important in counseling underachievers, particularly those who are using underachievement to rebel against, or even punish, the adults in their lives. These students need help in realizing that, in the long run, they themselves are the ones who will have to live with the outcomes of their underachievement. Their parents and teachers may experience unhappiness and frustration now, but ultimately will go on to live their own lives.

Implicit throughout all of the achievement theories is the concept of personal responsibility for outcomes—both immediate and long-range.

Summary. It should now be quite apparent that, as stated in the introduction, underachievement is a complex problem with both cognitive and affective dimensions caused by factors that are both internal and external to the student. In this section, many of these factors and their interactions have been identified. In the next section, suggestions are presented for making practical use of this information.

HOW CAN COUNSELORS INTERVENE?

In a 1967 ERIC document, *Guidance and the Underachiever* (ED 023 112), it was stated, "The research literature from 1963 through 1966 contains more than three times as many studies relating to the characteristics of underachievers as it does studies pertaining to procedures with which to aid them" (p. 10). Fortunately, we seem to have made considerable progress in the past twenty years. In fact, a review of the underachievement literature in the ERIC database from 1966 to 1986 revealed that over 80% of the articles and documents deal with program descriptions. Such descriptions, of course, do not constitute experimental research. Moreover, very few of the programs described have been formally evaluated by either internal or external observers. In the aggregate, however, even subjective observations of program outcomes can provide useful clues about the kinds of intervention strategies that seem to work. In this section, an analysis of this literature has been used to develop a seven-step plan for improving counseling interventions with underachievers.

Step 1. Redefine the Role of the School Counselor

"The school counselor should be first and foremost an educator, supporting students in their progress through the school and serving as a crucial resource for the academic program. Although counselors may spend much of their time providing direct services to students, both individually and in groups, they also can provide an important service to students indirectly by calling attention to conditions in schools and classrooms that are defeating or frustrating youngsters rather than helping them succeed. Through the nature of their work, counselors develop a sense of the crosscurrents in a school and the way they affect individual students. With the support of principals, they can lead a movement for more academic flexibility and more personal attention to the needs and problems of students. The role we are suggesting would make counselors key

. . . counselors . . . can provide an important service to students indirectly by calling attention to conditions in schools and classrooms that are defeating or frustrating youngsters rather than helping them succeed.

monitors of student progress, developers of student potential, and facilitators of counseling arrangements. Specifically, this role would involve counselors in the following tasks:

- Coordinating the school guidance plan.
- Developing strategies for school staff to use in reaching students whose academic performance has slipped, or who most need assistance in developing self-confidence and in thinking about the future.
- Serving as advocates for students who are at risk of not meeting new standards and requirements, and who may need a second chance at a college-bound program.
- Setting up information systems and developing the curriculum for group advising sessions about such things as college planning and course choices.
- Planning and carrying out projects to inform parents about school objectives and how they can support them.
- Training paraprofessionals and/or volunteers to perform special duties.
- Working with teachers and administrators to enhance their capacities in the areas of guidance and counseling to which they can contribute.
- Developing a bank of community resources to which youngsters with special problems can be referred.

"It must be emphasized that our recommendation is to *change* the position and role traditionally assigned to counselors within the school, not to add more responsibilities to their daily routine. As we heard repeatedly in our investigation, there are already 'too few counselors trying to do too much for too many.'

"A reconsideration of the role of the counselor within the school requires that principals, counselors, and teachers change long-standing relationships, and that they take new risks. If the counselor's role is to change in the direction recommended here, counselors will have to develop closer alliances with teachers and link their work with students more directly to classroom experience. Teachers and principals will have to acknowledge that counselors can make special contributions to improving student academic performance and can work with them to create a healthy climate for learning in the classroom—and in the school as a whole" (*Keeping the Options Open*, 1986, pp. 11-13).

These statements, taken from the final report of the Commission on Pre-College Guidance and Counseling, advocate a role for school counselors that would be particularly beneficial to underachievers. Obviously, this new role calls for a considerable shift from an emphasis on direct counseling to an emphasis on coordination and consultation. While this shift will not be welcomed by many counselors, it is probably inevitable, and school counselors may be well advised to become agents, rather than victims, of the changes that are likely to take place. Thus, school counselors should consider developing yet another area of expertise, i.e., change agentry.

A key phrase in the above statements is "with the support of principals." One of the primary conditions that must be present if attempts at change are to be successful is administrative support. This support "must be verbally expressed, in both spoken and written form. The change agent . . . must have a clear mandate to proceed that is recognized . . . by all others who will be involved. Highly important, also, is that provisions be made to support the change effort in the form of materials, equipment, in-service training experiences, allocation of time, and the development of a psychological climate that favors acceptance by students, staff, clients, employees, parents, and/or community" (Walz & Benjamin, 1977, p. 9).

Another condition Walz and Benjamin identify as essential to bringing about change is that those who will be involved in the change must feel the need for it. Although many of the major educational reform reports appear to ignore guidance and counseling, studies are emerging that are likely to sensitize educators to the critical role played by counselors. For example, Lee and Ekstom (in press), using data from the first and second follow-ups of *High School and Beyond*, found that "guidance counseling services appear to be unequally available to all public high school students" (p. 2). They point out that "the advice students receive on selecting a high school curriculum track or planning an appropriate course of study is likely to come both from home and from school." However, "students from families of lower SES, of minority status, and from small schools in rural areas are less likely to have access to guidance counseling for making these important decisions at the beginning of their high school careers." Thus, "it appears that students who may need such guidance the most, since they come from home environments where knowledge of the consequences of curricular choices is limited, are least likely to receive it in their schools" (p. 2). Reports such as this will very likely have an impact on future guidance and counseling programs. Whether this impact takes the form of a reshaping of the school counselor's role, guided by school counselors themselves, or severe criticism and a redefinition of the role by others will depend greatly on school counselors' willingness to take a proactive stance.

One way to convince parents, teachers, administrators, and students that school counselors can make direct contributions to students' academic development is to demonstrate it.

One way to convince parents, teachers, administrators, and students that school counselors can make direct contributions to students' academic development is to demonstrate it. Castagna and Codd (1984) did just this by collaborating with teachers to implement a seven-session study skills unit within the English curriculum for all high school freshmen. As a result, "the study skills program . . . proved to be an effective means for enhancing student academic success while simultaneously increasing counselor visibility and improving counselor-faculty relations" (p. 42).

Step 2. Utilize Research Findings to Update Intervention Strategies

In counseling, as in many disciplines, there is a notable gap between research and practice. Like other professionals, unless trained specifically in the use of a new tool or technique, counselors tend to rely primarily on the skills they learned in their pre-service training and the insights they've acquired through their own experiences.

Keeping up to date on relevant research may be more difficult for counselors than for other education professionals because their expertise draws not just from counseling research *per se*, but also from the research of other disciplines such as education, psychology, social work, and human resource development. Also, because of their extensive responsibilities and full schedules, they have little time to devote to their own professional development and renewal.

Counselor educators, on the other hand, are expected to keep abreast of developments both in research and in the field. Counselor educators, therefore, can serve as school counselors' link to information on research, and school counselors can serve as counselor educators' link to the field. At the professional association level, this can take the form of a special ASCA-ACES task force to develop counseling strategies for addressing student academic achievement problems. At the individual level, school counselors can contact their former (or nearby) counselor educators to explore possibilities for collaborative action research. Using principles that have already been identified through previous research, the school counselor-counselor educator team can design and test the application and/or expansion of these principles in specific settings. Principles to be tested could be drawn from the achievement motivation research discussed in the previous chapter, or from studies, such as the following, that have investigated the characteristics of effective interventions at the program level.

Bednar and Weinberg (1970) analyzed 23 studies that evaluated the effectiveness of various treatment programs for underachieving college students. They concluded that "the treatment programs associated with improved student academic performance were characterized as (a) structured rather than unstructured, (b) lengthy rather than brief, (c) counseling aimed at the dynamics of underachievement used in conjunction with an academic studies program, (d) having high levels of therapeutic conditions (empathy, warmth, and genuineness), and (e) appropriate to the needs of the students" (p. 1).

In a more recent review which focused on K-12 low and underachieving students, Wilson (1986) identified the following dimensions of effective treatment programs: (1) group versus individual counseling; (2) structured versus unstructured approach; (3) long versus short treatment; (4) volunteers versus non-volunteers; and (5) counseling with study skills. Because of the younger population studied, there was also some evidence that parental involvement was a significant factor in program effectiveness.

Another counseling technique that has been demonstrated to be effective is the use of goal setting conferences (Gaa, 1979). As McHolland (1971) states, "the general underachiever is often most helped by short term goal setting. He begins to see in a matter of days or weeks that he can do a lot if he will give himself the chance to try" (p. 4). Schunk (1984) demonstrated that goals combined with rewards are particularly effective in increasing both achievement and self-efficacy.

. . . goals combined with rewards are particularly effective in increasing both achievement and self-efficacy.

Rittenhouse, et al. (1984) suggest that the mediation of other factors may explain the failure of research to show the expected relationship between attribution beliefs and subsequent performance. They suggest having students develop action plans as a way of bridging the gap between belief and behavior. Fishbein and Ajzen (1975) have proposed a "belief-attitude-intention-behavior model" which emphasizes the importance of "intention" in predicting. Reinforcing this link by having students develop and commit themselves to written action plans can be a highly effective strategy for promoting follow-through behavior. A field test of the combination of these concepts (structured goal setting, action-planning, and rewards) would very likely yield quite useful information for counseling underachievers.

There has been very little research comparing the relative effectiveness of various counseling approaches in helping underachievers. Bednar and Weinberg's (1970) conclusions support a Rogerian approach. However, one study (Drury & Robbins, 1974) found significant differences favoring behaviorally trained counselors. This would be consistent with the fact that several projects using self-control training (e.g., Greiner & Karoly, 1976; Richards, 1975) have been able to demonstrate high levels of effectiveness. However, the subjects in the Drury and Robbins study were pre-delinquents and most of the self-control training studies have been done with college students and adults. Further research with students in K-12 settings would be very useful, particularly in terms of the implications for counselor education programs and the selection of counseling models for the pre-service training of counselors.

Step 3. Develop a Comprehensive Model for Intervention

Programs designed for underachievers can be classified into three types: primary strategies (direct study skills instruction), support strategies (counseling), and dual or multiple intervention approaches. Research results overwhelmingly support the third approach for both short- and long-term effectiveness. An organizing framework with the multiple intervention approach needs to be used to insure that the individual strategies provide comprehensive coverage and complement, rather than compete with, one another. The diagram on the following page illustrates a proposed model for organizing intervention strategies to deal with achievement problems. Although the model is still in the developmental stage, it can provide useful guidance for designing a comprehensive intervention program to meet the specific needs of individual underachievers.

Description of the Model. In his extensive work in the area of counselor development and renewal, Garry Walz frequently uses the term "empowerment" to describe the ultimate goal he has for persons attending his workshops. This term is particularly appropriate because it comprehensively integrates the various factors that must be present for participants to actually use the workshop outcomes. First, it suggests that the person feels powerful, confident, and capable of achieving his/her goals. Second, it suggests that the person knows what he/she must do and has the ability to do it. And, third, it suggests that the person is equipped with the prerequisite knowledge, skills, and resources necessary to implement his/her plan of action.

. . . empowerment serves as both the ultimate and the enabling goal for helping underachievers improve their academic performance.

In the proposed model, empowerment serves as both the ultimate and the enabling goal for helping underachievers improve their academic performance. It is similar to the concepts of perceived control and self-efficacy in the achievement motivation literature, but is somewhat broader in that it represents not only a feeling on the part of the student, but also the outcome of an action; i.e., the student becomes "empowered" through the achievement process.

Preceding empowerment are the more commonly identified outcomes of the achievement process—learning and grade point average. While there has been considerable debate regarding the relationship between these two outcomes, the issues involved are beyond the scope of the present discussion. Both have been included in the model because both are important.

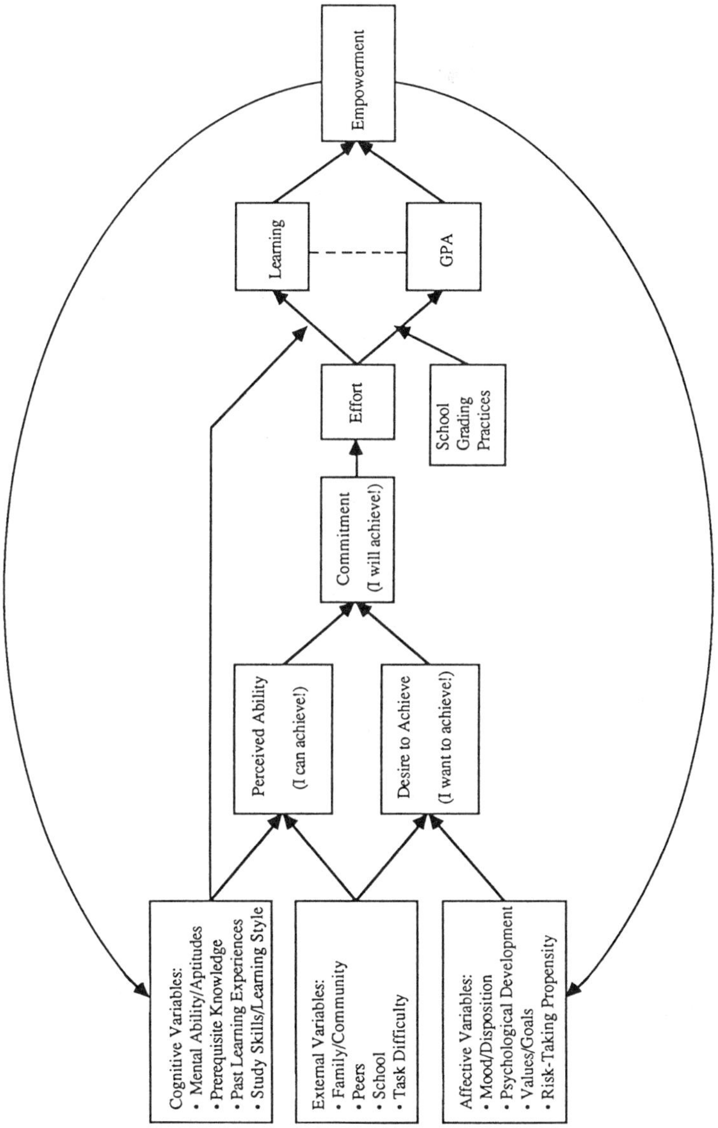
Empowerment
Learning
GPA
Effort
School Grading Practices
Commitment (I will achieve!)
Perceived Ability (I can achieve!)
Desire to Achieve (I want to achieve!)
Cognitive Variables:
• Mental Ability/Aptitudes
• Prerequisite Knowledge
• Past Learning Experiences
• Study Skills/Learning Style
External Variables:
• Family/Community
• Peers
• School
• Task Difficulty
Affective Variables:
• Mood/Disposition
• Psychological Development
• Values/Goals
• Risk-Taking Propensity

In order for learning to occur and grades to be earned, the student must put forth at least some amount of effort. The model attempts to show that the amount of effort needed for learning to occur can be modified by the student's cognitive abilities and previous experiences. Similarly, there may not be a direct correlation between effort and grades earned because of variability in teacher and school grading practices.

"Commitment" represents the point at which the student makes the decision to put forth effort toward achievement. It is the product of both the student's perception that he/she *can* achieve and the decision that he/she *wants* to achieve. This part of the model is based on Atkinson's expectancy-value theory. In the diagram, the two variables (perceived ability and desire) are given equal weight. However, it may be that a high level of ability may compensate for a low level of desire and vice versa. Further research is needed to determine the relative importance and potential interaction of these two variables.

At the far left of the model are the "input" variables. These are the abilities, attitudes, feelings, experiences, and external resources that the student brings to each achievement situation. As indicated in the model, the student's perceived ability is a function of both cognitive variables and external variables, while his/her desire to achieve is a function of both external variables and affective variables. It should be noted that further development of the model may lead to an alternative configuration that will provide for the possibility that the student's cognitive variables may affect desire to achieve, and his/her affective variables may affect perceived ability.

An important feature of this model is that the student's cognitive variables are viewed as consisting of not only his/her mental ability and specific aptitudes, but also such variables as subject-related prerequisite knowledge, past learning experiences, study skills, and learning style. The purpose for defining cognitive variables so broadly is to encourage counselors not to quickly accept the student's "ability" as a fixed entity, but rather to: (1) be sure that the student's actual ability has been fairly and accurately diagnosed considering his/her past experiences, socioeconomic factors, validity of the tests used, etc.; and (2) look for ways that the student's academic ability can be enhanced. For example, gaps in prerequisite knowledge can be addressed through tutoring or switching the student to a more basic course in the subject area. Past learning failures can be replaced by carefully programmed learning successes. Study skills can be improved through direct instruction and behavior modification techniques. The student can be helped to determine and capitalize on his/her optimum learning style. Even academic ability, per se, can be improved through the development of the student's critical thinking skills. In the past, academic ability was assumed to be directly related to mental ability, which, in turn, was assumed to be a relatively fixed trait. Now, however, there is an increasing acceptance of the idea that "intelligence" consists of various types of mental abilities and that "developed abilities" are a major factor both in test performance and in subsequent mental and academic performance.

. . . "intelligence" consists of various types of mental abilities and . . . "developed abilities" are a major factor both in test performance and in subsequent mental and academic performance.

Another important difference between this model and the achievement motivation theories that were reviewed is the weight given to external variables (family, peer, school, and community influences). In the achievement theories, task difficulty and luck are the external variables that receive most attention. Also, while the achievement motivation theories emphasize values and beliefs, the other affective variables—mood, psychological needs (e.g., adolescent reaction), and life/career goals—are addressed only implicitly.

Risk taking propensity has been addressed extensively in other areas of research in psychology, but little has been done to investigate its specific relevance to academic achievement. The rationale for including it in this model is that learning often involves change, and daring to undertake a new learning experience means having to risk the possibility of failure.

Use of the Model. In general, using the model to design an intervention strategy for an individual student involves assessing the student's status with respect to each of the input variables. Then, for each input variable that appears to be a barrier to the student's achievement, a strategy is devised to overcome the barrier. By using an integrated, multiple intervention approach to address all potential barriers, and by implementing strategies that reinforce the achievement process, counselors can significantly increase the likelihood of intervention effectiveness. For example, the use of behavior contracts can reinforce motivation, collaborative peer monitoring can reinforce effort, and celebration can reinforce empowerment.

The following procedure is suggested for using the proposed model:

1. Identify the student's assets and barriers by examining each of the twelve areas that comprise the "input" variables. This may require examination of the student's records, administration of specific diagnostic instruments (e.g., the Brown-Holtzman *Survey of Study Habits and Attitudes,* The Psychological Corporation) and interviews with parents, teachers, and the student. Appendix A ("Guidelines for Identifying Underachievement Factors") contains a set of questions that can be useful in collecting this information; counselors should add other questions that seem to be relevant.
2. Summarize your findings on the Student Achievement Assets and Barriers form provided in Appendix B, preferably in collaboration with the student. This not only allows the student to have input, but also provides an opportunity for the counselor to help the student acquire a realistic perception of his/her situation. Another advantage of a collaborative approach is the

student may be more likely to feel "ownership" of (thus, commitment to) the intervention process.

3. For each area that received a "1" or "2" rating, generate as many ways possible that the situation could be changed. Here, again, collaborative brainstorming with the student is likely to be more productive than trying to come up with ideas alone. Appendix C provides a format for recording these ideas.
4. For each "problem" area, select the change strategy that promises to be most effective and develop an action plan for implementing the strategy. This may require the involvement of significant others such as parents or teachers. If so, it is important they be included in the planning process. Appendix D provides a format for developing each implementation plan.
5. Summarize the set of action plans by developing a timeline to monitor progress. Here, the action steps from all of the action plans should be merged into one chronological list. Rewards should be granted as each goal is met. Appendix E provides one possible format for developing this type of timeline.

Although this process could be carried out through individual counseling, a group counseling approach is likely to be much more effective. For example, in Step 3, group interaction is likely to result in the generation of quite innovative and creative ideas. Also, the group setting is likely to provide students with a sense of peer understanding and support. It also encourages a higher level of accountability and follow-through on action plans.

Step 4. Create a Resource Bank of Ideas and Materials

In generating possible strategies for overcoming barriers (No. 3 in Step 3 above) and reinforcing the achievement process, counselors can contribute most if they have an extensive set of ideas and materials from which to draw. Such a "bank" should grow over time, ultimately filling an entire file drawer. The following set of suggestions, arranged to reflect the proposed intervention model, provides a "starter set."

Academic Ability. As discussed previously, academic ability is no longer viewed as a totally fixed trait of a student. This is not to say that all students can be expected to achieve at the same high level. What it does mean is that, because academic ability is highly related to previous learning experiences and developed abilities, many students, particularly long-time underachievers, can be helped to improve their current academic ability through direct instruction. In fact, many of the educational reform reports call for greater attention to the development of *all* students' critical thinking skills. This, in turn, has resulted in a rapidly growing body of literature on metacognition, problem solving skills, and strategies for helping students "learn how to learn." (See, for example, Derry & Murphy, 1986.) Counselors may find it very useful to collect some of these articles and documents for the purpose of creating their own strategies for helping students develop their learning skills. Unfortunately, much of the current literature is in the theoretical rather than the practical stage of development. Here is an area, however, where counselors can make a significant contribution to the educational program. First, the skills to be developed are ones that cut across all subject areas and thus are not the domain of an individual teacher. Each teacher is expected to emphasize critical thinking skills and effective learning strategies as they apply to his/her subject. Counselors, therefore, are in an ideal position to help students integrate these skills and see the commonalities across subjects. Second, the skills to be developed are very closely related to the kinds of problem solving and decision making skills which counselors teach students to use in life/career planning. Therefore, counselors already have a fair amount of

. . . many students, particularly long-time under-achievers, can be helped to improve their current academic ability through direct instruction.

expertise in helping students approach tasks in a systematic way.

In terms of specific resources, counselors may find it helpful to develop or collect a set of one-page student handouts that summarize step-by-step strategies for problem solving, decision making, approaching academic tasks, etc. The counselor may also wish to have on hand selected aptitude and achievement tests to help students identify specific areas of strength and weakness and to double-check areas that may have been mis-diagnosed in the past.

Specific Prerequisite Knowledge. Sometimes just getting caught up can be overwhelming to a student—particularly if underachievement has become chronic. Here collaboration with the student's teacher(s) can be useful in determining: (1) if catching up is a problem; (2) if so, where the gaps are in the student's knowledge; and (3) what arrangements can be made to get the student caught up. If the teacher cannot provide individualized remedial help, it may be necessary to arrange for a tutor. As discussed previously, a tutor who focuses more on key concepts and strategies for approaching the subject matter is preferable to one who simply coaches the student on specific facts and details. Since the strategies that are most likely to be relevant to overcoming this type of barrier involve primarily coordinating and referral, there are probably few material resources necessary other than a list of qualified tutors.

Past Learning Experiences. If, as attribution theory maintains, the way a student approaches new learning tasks depends greatly on what he/she feels are the causes of past successes and failures, this would be useful information for the counselor to obtain. Although the counselor could use standardized instruments for determining this, a more readily available approach would be to have students complete a "Report Card Analysis" on forms such as those suggested in Appendix F. The patterns of the student's responses can then be used to identify internal vs. external tendencies.

If successes tend to be attributed to external causes, the student probably needs help building confidence and claiming "ownership" of his/her ability. This would also be true if the student attributed failures to lack of ability when, according to aptitude tests, he/she does have the necessary ability. If failures are consistently attributed to external causes, the validity of the "excuses" should be explored. However, if the student is merely projecting blame, he/she may need help in taking more responsibility for outcomes.

Study Skills/Learning Style. Appendix G is an ERIC/CAPS Fact Sheet on "Counseling for Study Skills" (Frenza, 1983). It presents a comprehensive summary of the issues relevant to study skills and learning style and contains a particularly useful discussion of assessment instruments.

Two additional ERIC/CAPS publications that provide up-to-date information on academic counseling are Griggs (1985), *Counseling Students Through Their Individual Learning Styles*, and Price and Griggs (1985), *Counseling College Students Through Their Individual Learning Styles*.

Two excellent resources for study skills instruction are Brown and Holzman (1987), *A Guide to College Survival*, and Ellis (1985), *Becoming a Master Student.* Although both are designed for college students, many of the tips and guidelines work equally well for high school or even junior high school students. Both are in workbook format, making them excellent "textbooks" for group guidance. Alternatively, counselors can use them as their own sourcebooks, selecting and adapting the ideas for use with younger students. Appendix H contains an excerpt on notetaking from the Brown and Holzman book, and Appendix I contains an excerpt on taking multiple choice tests from the Ellis book.

Task Difficulty. In the achievement motivation literature, "tasks" are very specific assignments, such as solving a math problem or putting together a puzzle, where "difficulty" is a property of the task itself. In real school settings, it is probably more useful to think of "task difficulty" not only in terms of specific assignments, but also in a broader sense; i.e., the requirements that must be met to achieve a particular grade in a given course or subject. While there is often not much that counselors can do to change the nature of specific assignments or requirements, they can do the following: (a) through testing and/or interviews, help the teacher and/or parent assess the student's readiness to do the work; (b) help the teacher and/or student identify alternative learning strategies based on the student's learning style; and/or (c) help locate a tutor who can take the time to present the tasks in a different manner and/or at the student's own learning pace. Since these strategies are based primarily on consultation skills, it is difficult to identify specific material resources that counselors would need. Secondary school counselors may find it useful to collect and file specific course requirements and grading policies from each teacher as an aid in advising students on course selection and helping them appraise their readiness and commitment to take on difficult subjects and courses.

Family/Community Support. Probably the most effective way that parents can provide concrete help to underachieving students is through the monitoring of time spent on homework. It has been well documented that, for all ability levels, increased homework time results in higher achievement. In fact, one investigator found that low ability students spending only one to three hours on homework per week can achieve grades comparable to average ability students who do no homework (Keith, 1982). To help reinforce parents' involvement in monitoring homework, some counselors have found it useful to have students maintain a "homework log." This is simply a form which parents initial or sign when the student has completed each assignment or devoted a specified amount of time to studying.

Probably the most effective way that parents can provide concrete help to underachieving students is through the monitoring of time spent on homework.

Elementary and middle school counselors can have students design their own "Homework in Progress—Do Not Disturb" or "Danger—Homework Area—Do Not Enter" signs. This encourages students to take control

Secondary school counselors can play an extremely valuable role in serving as the liaison between the family and the school.

of setting their homework times and communicates to parents and siblings that they need to respect and support these efforts.

Secondary school counselors can play an extremely valuable role in serving as the liaison between the family and the school. In elementary school, students generally have only one teacher (except for special subjects like art, music, etc.); parent-teacher conferences are also regularly scheduled for children in elementary school. Therefore, parents at least know who their children's teachers are and can feel reasonably comfortable contacting their child's teacher to discuss any concerns they may have about their child's progress. However, when students enter junior high, they may have six different teachers every day, and parents usually visit the school only once or twice a year for an "open house." It is thus more difficult for parents to know whom to contact if their child is having problems in more than one subject. Also, as their children get older, parents tend to have fewer informal interactions with the parents of their children's friends. This makes it difficult for parents to present a "united front" in terms of expected study times, social activities, etc. Parent meetings at the school, organized and conducted by the school counselors, can help to meet both of these needs; i.e., they can increase the counselor's visibility as the school-home liaison and can provide a setting for informal interaction among parents themselves.

Peer Support. Kehayan (1983) has developed a program for underachievers called the "Peer Intervention Network (PIN)." In this group counseling approach, three types of peer resources are identified: experts, shadows, and consultants. "Experts" are identified early in the group process by having each member write down his/her own special areas of expertise (e.g., spelling, math, history, etc.) along with his/her name and phone number. Group members are then encouraged to call on one another for help with particular assignments. Each member then selects a "shadow" from the group who will help that member identify problem areas, set goals, and design a plan of action to accomplish his/her goals. The shadow then monitors the member's progress toward his/her goals (thus, the concept of "shadowing") and reports on the member's progress during group meetings. "Consultants" are identified later in the group process and consist of members who have met their goals relative to grade improvement and no longer require shadows. Results of a three-year study of this approach showed that group development moved from resistance to polarity to unity and that grade point averages improved from 1.0 to 2.0.

School Support. One way to promote students' feelings that the school is supportive of their personal development and achievement is through classroom guidance. Gerler and Anderson (1986) developed a 10-session classroom guidance unit titled "Succeeding in School" and tested it with fourth and fifth graders. Results of the study showed positive effects of the unit on classroom behavior, attitude toward school, and language grades. The ten sessions, which were 30 to 40 minutes each, provided opportunity for discussion of and, where appropriate, practice in the following concepts:

1. Success in school—ground rules for discussion, rationale for the unit.
2. Being comfortable in school—relaxation methods.
3. Being responsible in school—meaning, self-identification of examples of responsible behavior.
4. Listening in school—listening skills, importance and benefits of good listening.
5. Asking for help in school—games designed to improve students' skills in listening to teachers and asking them for help.
6. How to improve at school—identification of subjects in which students would like to

improve, how they might work toward improvement, examples of own past improvements.

7. Cooperating with peers at school—role playing of cooperation, discussion of benefits.
8. Cooperating with teachers—discussion of unfinished statements such as, "I would like to get along better with my teacher, but my problem is. . . ."
9. The bright side of school—identification and discussion of positive happenings at school; ways to change negative aspects.
10. The bright side of me—discussion of personal strengths, exchange of positive feedback with one another.

. . . sharing the guidance perspective with teachers is likely to result in increased understanding of and personal support to students.

Some of the educational reform reports have emphasized the guidance function as a teacher role rather than a school counselor role. Although this can be viewed and, in fact, may be intended as a threat to counselors, some counselors and counselor educators see it as an opportunity to highlight their expertise. Using the recommendations of the reform reports as a rationale for action, counselors, with the support of administrators, can organize and conduct formal meetings in which they share with teachers the ideas and resources they have found to be most helpful with various types of students and student problems. In this process of "giving away" counseling, counselors may actually increase their visibility and strengthen their position. More important, sharing the guidance perspective with teachers is likely to result in increased understanding of and personal support to students.

"Low regard for self or poor self-concept seems to be present in all underachievement no matter what else is involved.

Mood/Disposition. A student's mood or disposition, whether temporary or a general tendency, is strongly related to his/her self-concept and self-confidence. As McHolland (1971) points out, "Low regard for self or poor self-concept seems to be present in all underachievement no matter what else is involved. What it amounts to is that students achieve to their level of self-image and self-confidence" (p. 2). To improve underachievers' self-concept, and thus elevate their mood and help them develop a more positive disposition toward learning and achievement, McHolland recommends two group approaches. The first approach, the Human Potential Seminar, is designed for the general underachiever, i.e., the student who isn't using all of his/her potentiality, but would like to. The Human Potential Seminar focuses on: identification of the chief formative people in each group member's life; analysis of the individual's experiences of satisfaction, success, and achievement to identify his/her own unique motivators; acknowledgement of personal strengths; clarification of values; and pairing of values with short-term goal setting. According to McHolland (1971), "Through value clarification, a person begins to center in on what really matters to him—what his priorities are, and what he wants his life to be like. When we pair short-term goal setting with values, a person is helped to run his life responsibly and with purpose. He is determining his own life" (p. 4).

For the "hard core, committed underachiever," McHolland advocates the "Success Group." In this approach, group members are first asked to compile a list of "cop-outs" or excuses which they have used or could use to avoid achievement. These can range from serious (e.g., I was too sick to go to class) to outlandish (e.g., the dog ate my assignment). The main idea is to have the group identify as many excuses as possible to demonstrate "how expert they have been at *planning for failure*. The point of this is to show the students that they make decisions, they run their life—they are accountable for whether they succeed or fail" (p. 6). Students then make written commitments of their grade point achievement goals and specific short-term goals as to class attendance, participation, and homework. The goals are posted and reviewed each time the group meets. Cop-outs are challenged and progress is praised. In regard to feedback, McHolland points out, "as a student learns first to set short-term goals that are specifically measurable, achievable in a given time span, believable to him, and decisive, and then begins to achieve them, he provides himself with feedback. He can look at what his goal was and see whether he achieved it. He need not wait for weeks for feedback from a teacher or counselor. He begins to provide it for himself. Of course, the external feedback is reinforcing and helps but for the most part schools are not geared toward giving it" (p. 7).

Both the Human Potential Seminar and the Success Group approaches appear to have significant positive impacts on self-concept and academic achievement. These types of outcomes can very well set in motion an "upward spiral" in which a positive mood/disposition increases openness to learning and achievement, and achievement, in turn, creates a more positive mood/disposition on the part of the student.

Pobably the most common psychological barrier interfering with students' achievement is the dependence-independence conflict.

Psychological Needs. Probably the most common psychological barrier interfering with students' achievement is the dependence-independence conflict. One approach to helping students develop a positive and responsible attitude toward independence would be to use deCharms' (1968) "origin/pawn" concept to stimulate group discussion and promote more origin type attitudes and behavior. Examples of a discussion question might be, "Does doing what teachers (parents) want you to do make you a pawn?" or "At what age should people become origins?"

Values/Goals. There are many resources which counselors can use to help students clarify their values and set appropriate education and career goals, ranging from highly comprehensive and structured computer-assisted career guidance systems (e.g., SIGI and DISCOVER) to less structured group discussion techniques. Blackman (1971) used the Rokeach Value Survey (see Rokeach, 1968) as a discussion tool and found that the discussion of life values was a viable motivative force, with the results extending to both an improvement in grade point average and a decrease in disciplinary referrals. In essence, the Rokeach Value Survey calls for a ranking of the importance of:

(1) instrumental values, such as "ambitious," "courageous," "loving," "responsible," etc.; and (2) terminal values, such as "comfortable life," "sense of accomplishment," "world peace," and "pleasure." A complete listing of the sets of values and the group counseling strategies can be found in Blackman (1971). Alternatively, counselors can create their own lists, adjusting the terms used for the age level of the student. "Write-ins" by students could also be encouraged.

Another informal technique to help students become more aware of the values they already possess is to have them conduct their own "cost-benefit analysis" of achievement. This could be accomplished by having them list all the short-, intermediate-, and long-term benefits of achievement that they can think of on one sheet of paper and all the short-, intermediate-, and long-term costs of achievement on another sheet of paper. They should then review their lists and circle or asterisk those that are most important to them personally. These lists can then be analyzed, as a group or in dyads, to identify the number of costs vs. benefits, short- vs. long-term outcomes, and significant patterns in the nature of costs and benefits (e.g., social acceptance, parent approval, career goals, etc.).

Risk Taking Propensity. One creative approach to assessing and addressing students' risk taking propensity is through the use of computer games and simulations. By having small groups of students go through such programs as Lemonade Stand, Oregon Trail, Wizardry, and Desert Survival, the counselor can both observe and use as a basis for discussion examples of risk taking behaviors on the part of individual students. Under the guidance of both the counselor and other group members, students who are too reluctant to take risks can be encouraged to practice greater risk taking on subsequent games and simulations. Similarly, those who take too many risks can be encouraged to engage in more analysis and problem solving behavior before making decisions. The counselor can then help students see the relationship between their risk taking actions and

One creative approach to assessing and addressing students' risk taking propensity is through the use of computer games and simulations.

their achievement behavior, particularly as it relates to personal achievement goal setting and fear of failure.

Perceived Ability. The most comprehensive approach to reinforcing students' perceived ability (as well as their desire to achieve) is to help them consciously integrate the outcomes of all of the counseling activities that have been used to address the sixteen "input" variables described above. Appendix B (Student Achievement Assets and Barriers) can be used not only as an initial needs assessment instrument, but also as a tool for organizing and summarizing the insights that develop later on in the counseling process.

One strategy for increasing students' perceptions of their ability to achieve is to teach them time management skills. As suggested by the title of a book by one famous time management expert (Lakein, 1973), getting control of your time means getting control of your life. For high school students, counselors can use any of the many popular paperbacks on time management strategies; for elementary and junior high school students, they can extract the principles and translate them into terms more relevant for students' roles, responsibilities, and potential distractors. Alternatively, counselors can draw from the time management strategies developed specifically for students by Brown and Holtzman (1987) and/or by Ellis (1985). Ellis suggests that students first analyze how they are currently spending their time by using a "Time Monitor" form for one week and, at the end of the week, completing the following statements (p. 42):

1. I want to spend more time on . . .
2. I want to spend less time on . . .
3. I was surprised that I spent so much time on . . .
4. I was surprised that I spent so little time on . . .
5. I had strong feelings about . . .

Ellis then goes on to provide "21 ways to get the most out of NOW"—time management techniques about "when to study, where to study, how to handle the rest of the world, and things you can ask yourself when you get stuck" (p. 47). These suggestions include such things as: study difficult (or boring) subjects first; avoid scheduling marathon study sessions; be aware of your best time of day; notice how others misuse your time.

Brown and Holtzman (1987) also provide helpful guidelines for student time management and discuss a five-step sequence for preparing a "Time Budget." This includes scheduling time for fixed commitments, daily living activities, course preparation, course review, and recreation. One of the more useful suggestions is to specify the time to study *each course* rather than just setting aside general "study time."

Desire to Achieve. Ongoing counseling support and positive feedback from significant others are important factors in the reinforcement and maintenance of students' desire to achieve. One strategy for directly enhancing this desire is to match students with local role models who are in the jobs or careers that the students have selected through career guidance activities. Drawing from parents as well as other community members, counselors can create a resource pool of adults who would be willing to meet with individual or small groups of students (preferably in their own workplace) to help the students get a better idea of what their jobs are really like. It would be especially important for these people to describe the linkages between what students are currently learning in school and what they will find useful in solving problems on the job. For example, an architect might select a roof design where, to obtain the exact measurements desired, a basic geometry theorem can be applied. Similarly, a social worker or counselor might be able to show how knowledge of a foreign language was particularly helpful in communicating with a recent immigrant or foreign student.

Ongoing counseling support and positive feedback from significant others are important factors in the reinforcement and maintenance of students' desire to achieve.

Commitment. Counselors can help students commit themselves to engaging in achievement-oriented behavior by collaboratively developing behavior contracts that contain very specific statements. For example, "I agree to study algebra every night, Sunday through Thursday, from 7:00 to 7:30 with the understanding that only legitimate emergencies will be accepted for exceptions and all 'cop-out' excuses will be challenged." Appendix D (Action Plan for Overcoming Achievement Barrier) can also be useful in encouraging students to specify their commitment. McHolland (1971) points out that making these commitments public, i.e., keeping them posted for all group members to see, is a very effective way to reinforce determination and encourage accountability.

Effort. Strategies for reinforcing and helping students maintain achievement-directed effort include short-term goal setting and rewards and ongoing monitoring of such behaviors as class attendance and homework completion. Involving peers in these activities, as suggested by McHolland (1971) and Kehayan (1983), is particularly effective.

Learning/Achievement and Grade Point Average. A change in grade point average is obviously the achievement outcome that is most easily observed and measured. It is important, however, that students, teachers, parents, and counselors be realistic in their expectations. When the expected increase isn't achieved, the potential reasons should be explored. Perhaps it was already too late in the semester for a change in behavior to have an impact on that semester's grades. Or, perhaps while the student was focusing on improving in one very difficult subject area, he/she let achievement in other subject areas slip. In any case, it is just as important to help students analyze the causes for negative results as it is to help them reinforce their perceptions of and celebrate all positive results.

. . . it is just as important to help students analyze the causes for negative results as it is to help them reinforce their perceptions of and celebrate all positive results.

Aside from the observable GPA, counselors can help students gain an appreciation for their actual learning and achievement by having them reflect on what they know now compared to what they knew at the beginning of the semester for each course they took and, further, how they might be able to use their newly acquired skills and knowledge in future courses, in their chosen careers, and/or in their personal lives.

Empowerment. On the basis of the definition discussed earlier, empowerment is a feeling that students will experience if they have been successful in increasing their academic achievement. Therefore, it probably needs no reinforcement. Since it is likely to carry with it a feeling of exuberance and a desire to celebrate, it is also an excellent time to grant rewards and help students consciously solidify the linkages between achievement behaviors and the accompanying good feelings.

Step 5. Evaluate and Refine Intervention Strategies

Program evaluation and accountability have received a great deal of attention in the past twenty years. They are particularly difficult areas to address for counseling and guidance programs because program outcomes are both hard to measure and easily affected by factors external to the counseling program. Nevertheless, administrators do demand evidence of accountability from counselors, and counselors themselves like to know what impact their efforts are having on students' growth and development.

Perhaps more important than the summative evaluation that focuses on whether or not a program works is a formative evaluation that focuses on why a program works or doesn't work. Thus, summative evaluation is aimed primarily toward program accountability, while formative evaluation is aimed more at program improvement. If time and resources are available, it is highly desirable to maintain a formal, ongoing system of formative evaluation that utilizes one or more of the many sophisticated evaluation models that have been developed. However, even without such resources or expertise, counselors themselves can do a great deal to improve and refine their intervention strategies by scheduling regular periods of time to reflect on and document what they have done, their impressions of the outcomes, their insights regarding why some things worked and others didn't, why certain strategies seemed to work better with some students than with others, and what they would like to change to improve their programs. Here, the important ingredients are regular review and written observations. Many times counselors will have sudden insights and will make a mental note to try a new approach for a particular situation, but will forget about it until it's too late to implement before the situation arises again. Just as it is important for students to continually analyze, monitor, and evaluate the effectiveness of their achievement behaviors so that they can make necessary adjustments, it is important for counselors to do the same with their counseling strategies. By so doing, they not only provide themselves feedback on their own accomplishments, but also improve the services they can provide to students in the future.

. . . most counselors, particularly school counselors, are so busy doing their jobs that they have little time to document their successes and to share them with others. . . .

Step 6. Share Exemplary Practices

One of the most difficult problems facing the counseling profession is that most counselors, particularly school counselors, are so busy doing their jobs that they have little time to document their successes and to share them with others, either through written descriptions or presentations at professional conferences. For most school counselors, it is difficult to find the time to submit a presentation proposal, let alone prepare the actual presentation. Moreover, funds to support travel to professional conferences or time for professional writing tend to be extremely limited for school counselors. Therefore, most of the sharing of innovative ideas and effective, practical strategies takes place on a local, informal, word-of-mouth basis, and the profession as a whole is deprived of its own collective wisdom.

Needless to say, counselors should use whatever change agent skills they have to get administrators and supervisors to place a higher priority on, and thus provide the necessary released time and financial support for professional writing and conference presentations. In the meantime, counselors can use the "practice file" component of the ERIC system to share ideas with one another.

Practice-oriented materials are documents that have been screened and tagged specifically for practitioners. They include a variety of useful resources—for example, curriculum and training guides, fact sheets, and research summaries. Counselors are encouraged to contact ERIC/CAPS for information on developing materials for submission to ERIC, including how to write about practices or programs in a way that is most useful to practitioners.

BIBLIOGRAPHY

Arkava, M. L. (1969). Alterations in achievement motivation through counseling intervention. *Journal of Secondary Education, 44*(2), 74-80.

Atkinson, J. W. (1964). *An introduction to motivation.* Princeton, NJ: Van Nostrand.

Bandura, A. (1977). Self-efficacy: Toward a unifying theory of behavioral change. *Psychological Review, 84*, 191-215.

Bandura, A., Adams, N. E., & Beyer, J. (1977). Cognitive processes mediating behavioral change. *Journal of Personality and Social Psychology, 35*(3), 125-139.

Bar-Tal, D. (1978). Attributional analysis of achievement-related behavior. *Review of Educational Research, 48*(2), 259-271.

Bednar, R. L., & Weinberg, S. L. (1970). Ingredients of successful treatment programs for underachievers. *Journal of Counseling Psychology, 17*(1), 1-7.

Beers, P. G. (1986). Accelerated reading for high school students. *Journal of Reading, 29*(4), 311-315.

Benson, J., et al. (1986). Effects of test-wiseness training and ethnicity on achievement of third- and fifth-grade students. *Measurement and Evaluation in Counseling and Development, 18*(4), 154-162.

Blackman, E. L. (1971, April). *The motivational effect of values as content in group counseling with underachieving adolescents.* Paper presented at the Annual Meeting of the California Educational Research Association, San Diego, CA. (ERIC Document Reproduction Service No. ED 049 493)

Blaha, J. (1982). Predicting reading and arithmetic achievement with measures of reading attitudes and cognitive styles. *Perceptual and Motor Skills, 55*(1), 107-114.

Blanchard, J. S. (1985). What to tell students about underlining . . . and why. *Journal of Reading, 29*(3), 199-203.

Brown, R. T. (1980, September). *Some misgivings about locus of control orientation and its relationship to intelligence, academic achievement, and delinquency.* Paper presented at the Annual Convention of the American Psychological Association, Montreal, Canada. (ERIC Document Reproduction Service No. ED 197 263)

Brown, W. F., & Holtzman, W. H. (1987). *A guide to college survival, revised edition.* Iowa City, IA: The American College Testing Program.

Caliste, E. R. (1980). *The effect of counseling on the behavior of sixth grade underachievers.* Unpublished manuscript. (ERIC Document Reproduction Service No. ED 194 823)

Calsyn, R. (1973). *The causal relationship between self-esteem, locus of control and achievement: A cross-lagged panel analysis.* Unpublished doctoral dissertation. Northwestern University, Evanston, IL.

Calsyn, R. J., & Kenny, D. A. (1977). Self-concept of ability and perceived education of others: Cause or effect of academic achievement. *Journal of Educational Psychology, 69*, 136-145.

Campanile, P. (1981). Evening up the score. *Instructor, 90*(8), 58-59.

Castagna, S. A., & Codd, J. M. (1984). High school study skills: Reasons and techniques for counseling involvement. *School Counselor, 32*(1), 37-42.

Cauley, K. M., & Murray, F. B. (1982). Structure of children's reasoning about attributes of school success and failure. *American Educational Research Journal, 19*(3), 473-480.

Chaika, G. (1985). Teaching test-taking techniques to the gifted. *Clearinghouse, 59*(4), 182-184.

Christen, W. L., & Murphy, T. J. (1985). Learning how to learn: How important are study skills? *NASSP Bulletin, 69*(483), 82-88.

Cole, C. (1986). Obstacles to excellence in counseling. *The School Counselor, 34*(2), 85-86.

Connolly, S. G. (1980). Changing expectancies: A counseling model based on locus of control. *Personnel and Guidance Journal, 59*(3), 176-180.

Crandall, V. C., Katkovsky, W., & Crandall, V. J. (1965). Children's beliefs in their own control of reinforcement in intellectual academic achievement situations. *Child Development, 36*, 91-109.

Criscuolo, N. P. (1986). Parent involvement in the testing program. *Clearinghouse, 59*(7), 330-331.

deCharms, R. (1968). *Personal causation.* New York: Academic Press.

Derry, S. J., & Murphy, D. A. (1986). Designing systems that train learning ability: From theory to practice. *Review of Educational Research, 56*(1), 1-39.

Druian, G. (1986). *Effective schooling and at-risk youth: What the research shows.* Portland, OR: Northwest Regional Educational Laboratory.

Drury, R. L., & Robbins, C. A. (1974, April). *Behavioral outcomes for predelinquents receiving behavioral, generic, or no counseling.* Paper presented at the Annual Meeting of the Western Psychological Association, San Francisco, CA. (ERIC Document Reproduction Service No. ED 106 676)

Dudley-Marling, C. C., et al. (1982). Locus of control and learning disabilities: A review and discussion. *Perceptual and Motor Skills, 54*(2), 503-514.

Dweck, C. S. (1975). Role of expectations and attributions in the alleviation of learned helplessness. *Journal of Personality and Social Psychology, 31*, 674-685.

Dweck, C. S., & Reppucci, N. D. (1973). Learned helplessness and reinforcement responsibility in children. *Journal of Personality and Social Psychology, 25*, 109-116.

Eisen, S. V. (1979). Actor-observer differences in information inference and causal attribution. *Journal of Personality and Social Psychology, 37*, 261-272.

Ellis, D. B. (1985). *Becoming a master student.* Rapid City, SD: College Survival, Inc.

Entwistle, N. J., & Kozeki, B. (1985). Relationships between school motivation, approaches to studying, and attainment among British and Hungarian adolescents. *British Journal of Educational Psychology, 55*(2), 124-137.

Fenn, L. M., & Iwanicki, E. F. (1983, April). *An investigation of the relationship between student affective characteristics and student achievement within more and less effective school settings.* Paper presented at the Annual Meeting of the American Educational Research Association, Montreal, Canada. (ERIC Document Reproduction Service No. ED 233 101)

Fish, M. C., & Mendola, L. R. (1986). The effect of self-instruction training on homework completion in an elementary special education class. *School Psychology Review, 15*(2), 268-276.

Fishbein, M., & Ajzen, I. (1975). *Belief, attitude, intention and behavior: An introduction to theory and research.* Reading, MA: Addison-Wesley.

Frenza, M. (1983). *Counseling for study skills.* Ann Arbor, MI: The University of Michigan, ERIC Clearinghouse on Counseling and Personnel Services.

Gaa, J. P. (1979). The effect of individual goal-setting conferences on academic achievement and modification of locus of control orientation. *Psychology in the Schools, 16*(4), 591-596.

Gadwa, K., & Griggs, S. A. (1985). The school dropout: Implications for counselors. *The School Counselor, 33*(1), 9-17.

Gadzella, B. M. (1985). Research brief: Study skills presented through computer-assisted instruction to high school students. *Texas Tech Journal of Education, 12*(1), 47-50.

Gama, E. M. P., & de Jesus, D. M. (1986, April). *Teachers' expectations and causal attributions for students' academic achievement.* Paper presented at the Annual Meeting of the American Educational Research Association, San Francisco, CA. (ERIC Document Reproduction Service No. ED 272 281)

Gerler, E. R., Jr., & Anderson, R. F. (1986). The effects of classroom guidance on children's success in school. *Journal of Counseling and Development, 65*(2), 78-81.

Gifford, C. S., & Fluitt, J. L. (1980). How to make your students test-wise: Practice, practice practice. *American School Board Journal, 167*(10), 29, 40.

Glasser, W. (1975). *Reality therapy.* New York: Harper & Row.

Goodstein, M. (1980). The diagnosis and treatment of underachievement. *Journal of the International Association of Pupil Personnel Workers, 24*(2), 102-109.

Green, K. R., & Baker, A. (1986). *Promising practices for high-risk youth in the northwest region: Initial search.* Portland, OR: Northwest Regional Educational Laboratory.

Greiner, J. M., & Karoly, P. (1976). Effects of self-control training on study activity and academic performance: An analysis of self-monitoring, self-reward, and systematic-planning components. *Journal of Counseling Psychology, 23*(6), 495-502.

Griffore, R. J., & Parsons, M. A. (1983). Student characteristics and achievement in desegregated schools. *Urban Education, 17*(4), 431-438.

Griggs, S. A. (1985). *Counseling students through their individual learning styles.* Ann Arbor, MI: The University of Michigan, ERIC Clearinghouse on Counseling and Personnel Services.

Guidance and the underachiever. (1967). Albany, NY: State University of New York. (ERIC Document Reproduction Service No. ED 023 112)

Gunston-Parks, C., & Thomas, K. J. (1986). Writing as a study skill in the classroom learning spiral. *Reading Horizons*, *26*(3), 209-215.

Haines, P., McGrath, P., & Pirot, M. (1980). Expectations and persistence: An experimental comparison of Bandura and Rotter. *Social Behavior and Personality, 8*(2), 193-201.

Hamachek, D. E. (1972, April). *Effect of early school failure experiences on self-image development and implications for school counselors.* Paper presented at the Annual Meeting of the American Educational Research Association, Chicago, IL. (ERIC Document Reproduction Service No. ED 062 669)

Hess, G. A., et al. (1986). *"Where's Room 185?" How schools can reduce their dropout problem.* Chicago, IL: Chicago Panel on Public School Policy and Finance.

Hidi, S., & Klaiman, R. (1983). Notetaking by experts and novices: An attempt to identify teachable strategies. *Curriculum Inquiry*, *13*(4), 377-395.

Hillman, S. J. (1984, April). *Contributions to achievement: The role of expectations and self-efficacy in students, teachers, and principals.* Paper presented at the Annual Meeting of the American Educational Research Association, New Orleans, LA. (ERIC Document Reproduction Service No. ED 247 290)

Hodgkinson, H. L. (1985). *All one system: Demographics of education, kindergarten through graduate school.* Washington, DC: Institute for Educational Leadership, Inc.

Holbrook, H. T. (1986). ERIC/RCS report: Metacomprehension. *Journal of Reading*, *29*(6), 556-558.

Keeping the options open: Recommendations. Final report of the commission on precollege guidance and counseling. (1986). New York: College Entrance Examination Board. (ERIC Document Reproduction Service No. ED 275 948)

Kehayan, V. A. (1983, March). *Peer intervention network: A program for underachievers.* Paper presented at the Annual Convention of the American Personnel and Guidance Association, Washington, DC. (ERIC Document Reproduction Service No. ED 237 671)

Keith, T. Z. (1982). Time spent on homework and high school grades: A large-sample path analysis. *Journal of Educational Psychology, 74*(2), 248-253.

Keith, T. Z., Pottebaum, S. M., & Eberhart, S. (1985, April). *Effects of self-concept and locus of control on achievement.* Paper presented at the Annual Meeting of the National Association of School Psychologists, Las Vegas, NV. (ERIC Document Reproduction Service No. ED 260 327)

Kennelly, K. J., & Mount, S. A. (1985). Perceived contingency of reinforcements, helplessness, locus of control, and academic performance. *Psychology in the Schools, 22*, 465-469.

Kifer, E. (1975). Relationship between academic achievement and personality characteristics: A quasi-longitudinal study. *American Educational Research Journal, 12*, 191-210.

Kirschenbaum, D. S., & Perri, M. G. (1982). Improving academic competence in adults: A review of recent research. *Journal of Counseling Psychology, 29*(1), 76-94.

Kun, A. (1977). Development of magnitude-covariation and compensation schemata in ability and effort attributions of performance. *Child Development, 48*, 862-873.

Lakein, A. (1973). *How to get control of your time and your life.* New York: Signet.

Lee, V. E. *Multi-level models for social class and achievement.* (1986, April). Paper presented at the Annual Meeting of the American Educational Research Association, San Francisco, CA. (ERIC Document Reproduction Service No. ED 270 493)

Lee, V. E., & Ekstrom, R. B. (in press). *Student access to guidance counseling in high school.*

Licht, B. G. (1983). Cognitive-motivational factors that contribute to the achievement of learning-disabled children. *Journal of Learning Disabilities, 16*(8), 483-490.

Little, A. W. (1985). The child's understanding of the causes of academic success and failure: A case study of British schoolchildren. *British Journal of Educational Psychology, 55*(1), 11-23.

Maier, S., & Seligman, M. (1976). Learned helplessness: Theory and evidence. *Journal of Experimental Psychology: General, 105*, 3-46.

Markle, G. (1981). Improving test-taking skills of LD adolescents. *Academic Therapy, 16*(3), 333-342.

Marsh, H. W. (1984). Relations among dimensions of self-attribution, dimensions of self-concept, and academic achievements. *Journal of Educational Psychology, 76*(6), 1291-1308.

Martin, J., Marx, R. W., & Martin, E. W. (1980). Instructional counseling for chronic underachievers. *The School Counselor, 28*(2), 109-118.

McAndrew, D. A. (1983). Underlining and notetaking: Some suggestions from research. *Journal of Reading, 27*(2), 103-108.

McDill, E. L., Natriello, G., & Pallas, A. M. (1985). Raising standards and retaining students: The impact of the reform recommendations on potential dropouts. *Review of Educational Research, 55*(4), 415-433.

McGuire, D. E., & Lyons, J. S. (1985). A trans-contextual model for intervention with problems of school underachievement. *American Journal of Family Therapy, 13*(3), 37-45.

McHolland, J. D. (1971, February). *Human potential seminars: An approach to turning on the gifted underachiever.* Paper presented at the Annual Convention of the American Association of School Administrators, Atlantic City, NJ. (ERIC Document Reproduction Service No. ED 047 349)

Middleton, J. L. (1985). "I just can't take tests!" *Science Teacher, 52*(2), 34-35.

Miller, S. K., & Crano, W. D. (1980, April). *Raising low-income/minority achievement by reducing student sense of academic futility: The underlying theoretical commonalities of suggested strategies.* Paper presented at the Annual Meeting of the American Educational Research Association, Boston, MA. (ERIC Document Reproduction Service No. ED 186 575)

Myers, E. J. Counseling the parents of sixth grade underachievers. *Journal of Education, 154*(1), 50-53.

Natriello, G. (Ed.). (1986). *School dropouts: Patterns and policies.* New York: Columbia University, Teachers College.

Natriello, G., & McDill, E. L. (1986). Performance standards, student effort on homework, and academic achievement. *Sociology of Education, 59*(1), 18-31.

Nowicki, S., Jr. (1982). Competition-cooperation as a mediator of locus of control and achievement. *Journal of Research in Personality, 16*(2), 157-164.

Osborn, J. (1986). Reading comprehension. *Curriculum Review, 25*(5-6), 9-10, 12, 14, 16-26.

Parrish, B. W. (1982). A test to test test-wiseness. *Journal of Reading, 25*(7), 672-675.

Parsons, J. E. (1981, April). *Attributions, learned helplessness and sex differences in achievement.* Expanded version of paper presented at the Annual Meeting of the American Educational Research Association, Los Angeles, CA. (ERIC Document Reproduction Service No. ED 208 020)

Peck, N., Law, A., & Mills, R. C. (1987). *Dropout prevention: What we have learned.* Ann Arbor, MI: The University of Michigan, ERIC Clearinghouse on Counseling and Personnel Services.

Pettigrew, T. F. (1979). The ultimate attribution error: Extending Allport's cognitive analysis of prejudice. *Personality and Social Psychology Bulletin, 5*, 461-476.

Pigott, H. E., Fantuzzo, J. W., & Clement, P. W. (1986). The effects of reciprocal peer tutoring and group contingencies of the academic performance of elementary school children. *Journal of Applied Behavior Analysis, 19*(1), 93-98.

Poindexter, C. A., & Prescott, S. (1986). A technique for teaching students to draw inferences from text. *Reading Teacher, 39*(9), 908-911.

Price, G. E., & Griggs, S. A. (1985). *Counseling college students through their individual learning styles.* Ann Arbor, MI: The University of Michigan, ERIC Clearinghouse on Counseling and Personnel Services.

Reynolds, D. J. (1971). *Evaluating and extending learning counseling. Final report.* Washington, DC: National Center for Educational Research and Development (DHEW/OE). (ERIC Document Reproduction Service No. ED 057 414)

Richards, C. S. (1975). Behavior modification of studying through study skills advice and self-control procedures. *Journal of Counseling Psychology, 22*(5), 431-436.

Rittenhouse, J., Stephan, W. G., & LeVine, E. (1984). Peer attributions and action plans for under-achievement: Implications for peer counseling. *Personnel and Guidance Journal, 62*(7), 391-397.

Ritter, S., & Idol-Maestas, L. (1986). Teaching middle school students to use a test-taking strategy. *Journal of Educational Research, 79*(6), 350-357.

Robyak, J. E. (1977). A revised study skills model: Do some of them practice what we teach? *Personnel and Guidance Journal, 56*(3), 171-175.

Rokeach, M. (1968). A theory of organization and change within value-attitude systems. *Journal of Social Issues, 24*, 13-33.

Ross, L. (1977). The intuitive psychologist and his shortcomings: Distortions in the attribution process. In L. Berkowitz (Ed.), *Advances in experimental social psychology* (Vol. 10, pp. 173-220). New York: Academic Press.

Roth, R., & Meyersburg, H. (1963). The non-achievement syndrome. *Personnel and Guidance Journal, 41*, 535-540.

Rothbaum, F., Weisz, J. R., & Snyder, S. S. (1982). Changing the world and changing the self: A two-process model of perceived control. *Journal of Personality and Social Psychology, 42*(1), 5-37.

Rotter, J. B. (1966). Generalized expectancies for internal versus external control of reinforcement. *Psychological Monographs, 1* (Whole No. 609).

Rotter, J. B. (1975). Some problems and misconceptions related to the construct of internal versus external control of reinforcement. *Journal of Consulting and Clinical Psychology, 43*, 56-67.

Sadowski, C. J., & Woodward, H. R. (1981, April). *Teacher, locus of control and students' perceptions and performance.* Paper presented at the Annual Convention of the American Psychological Association, Los Angeles, CA. (ERIC Document Reproduction Service No. ED 209 591)

Sattes, B. D. (1985). *Parent involvement: A review of the literature.* AEL Occasional Paper 021. Charleston, WV: Appalachia Educational Laboratory.

Schunk, D. H. (1981, April). *Effort attribution: The direction makes a difference.* Paper presented at the Annual Meeting of the American Educational

Research Association, Los Angeles, CA. (ERIC Document Reproduction Service No. ED 199 284)

Schunk, D. H. (1981). Modeling and attributional effects on children's achievement: A self-efficacy analysis. *Journal of Educational Psychology*, *73*(1), 93-105.

Schunk, D. H. (1983). Ability versus effort attribution feedback: Differential effects on self-efficacy and achievement. *Journal of Educational Psychology*, *75*(6), 848-856.

Schunk, D. H. (1983, April). *Self-efficacy enhancement through motivational and informational processes*. Paper presented at the Annual Meeting of the American Educational Research Association, Montreal, Canada. (ERIC Document Reproduction Service No. ED 227 167)

Schunk, D. H. (1984). Enhancing self-efficacy and achievement through rewards and goals: Motivational and informational effects. *Journal of Educational Research*, *78*(1), 29-34.

Schwab, P. N., & Coble, C. R. (1985). Reading, thinking, and semantic webbing. *Science Teacher*, *52*(5), 68-71.

Shilling, F. C. (1984). Teaching study skills in the intermediate grades--we can do more. *Journal of Reading*, *27*(7), 620-623.

Shoop, M. (1986). InQuest: A listening and reading comprehension strategy. *Reading Teacher*, *39*(7), 670-674.

Simpson, M. L. (1986). PORPE: A writing strategy for studying and learning in the content areas. *Journal of Reading*, *29*(5), 407-414.

Smith, D. L. (1976). Goal attainment scaling as an adjunct to counseling. *Journal of Counseling Psychology*, *23*(1), 22-27.

Smith, R. L., & Troth, W. A. (1975). Achievement motivation: A rational approach to psychological education. *Journal of Counseling Psychology*, *22*(6), 500-504.

Solomon, L. J., & Rothblum, E. D. (1984). Academic procrastination: Frequency and cognitive-behavioral correlates. *Journal of Counseling Psychology, 31*(4), 503-509.

Stipek, D. (1980). A causal analysis of the relationship between locus of control and academic achievement in first grade. *Contemporary Educational Psychology, 5*, 90-99.

Stipek, D. J., & Weisz, J. R. (1981). Perceived personal control and academic achievement. *Review of Educational Research, 51*(1), 101-137.

Tesiny, E. P., et al. (1980). Childhood depression, locus of control, and school achievement. *Journal of Educational Psychology, 72*(4), 506-510.

The West Virginia Dropout Study: 1985-1986. (1987). Charleston, WV: West Virginia Department of Education.

Thomas, J. W. (1980). Agency and achievement: Self-management and self-regard. *Review of Educational Research, 50*(2), 213-240.

Thomas, T. (1985). *The Oakland County attendance/dropout task force: Who we are—where we're going*! Pontiac, MI: Oakland Schools.

Turvey, J. S. (1986). Homework—Its importance to student achievement. *NASSP Bulletin, 70*(487), 27-35.

Vargas, M. F. (1985). Developing an immunity to sophomoric plagiarism: Notetaking skills. *English Journal, 74*(2), 42-44.

Walz, G. R. (1984). *Counseling and educational excellence: A response to a nation at risk.* Ann Arbor, MI: The University of Michigan, ERIC Clearinghouse on Counseling and Personnel Services.

Walz, G. R., & Benjamin, L. (1977). *On becoming a change agent.* Ann Arbor, MI: The University of Michigan, ERIC Clearinghouse on Counseling and Personnel Services.

Weiner, B. (1979). A theory of motivation for some classroom experiences. *Journal of Educational Psychology*, *71*, 3-25.

Weiner, B., Frieze, I., Kukla, A., Reed, L., Rest, S., & Rosenbaum, R. M. (1971). Perceiving the causes of success and failure. In E. E. Jones, et al. (Eds.), *Attribution: Perceiving the causes of behavior*. Morristown, NJ: General Learning Press.

Whimbey, A. (1985). Reading, writing, reasoning linked in testing and training. *Journal of Reading*, *29*(2), 118-122.

Whitmore, J. R. (1986). Understanding a lack of motivation to excel. *Gifted Child Quarterly*, *30*(2), 66-69.

Willig, A. C., Harnisch, D. L., Hill, K. T., & Maehr, M. L. (1983). Sociocultural and educational correlates of success-failure attributions and evaluation anxiety in the school setting for Black, Hispanic, and Anglo children. *American Educational Research Journal*, *20*(3), 385-410.

Willis, H. D. (1986). *Students at risk: A review of conditions, circumstances, indicators, and educational implications*. Elmhurst, IL: North Central Regional Educational Laboratory.

Wilson, N. S. (1986). Counselor interventions with low-achieving and underachieving elementary, middle, and high school students: A review of the literature. *Journal of Counseling and Development*, *64*(10), 628-634.

Wilson, N. S. (1986). Effects of a classroom guidance unit on sixth graders' examination performance. *Journal of Humanistic Education and Development*, *25*(2), 70-79.

Wilson, N. S. (1986). Preparing for examinations: A classroom guidance unit. *School Counselor*, *33*(4), 297-305.

Wolf, F. M., et al. (1980, March). *Quality of school life: Cause or effect of beliefs of academic responsibility? A cross-lagged panel correlation analysis*. Paper presented at the Annual Meeting of the Eastern Educational Research Association, Norfolk, VA. (ERIC Document Reproduction Service No. ED 194 537)

Wolf, F. M., & Savickas, M. L. (1985). Time perspectives and causal attributions for achievement. *Journal of Educational Psychology, 77*(4), 471-480.

Zuccone, C. F., & Amerikaner, M. (1986). Counseling gifted underachievers: A family systems approach. *Journal of Counseling and Development, 64*(9), 590-592.

APPENDIX A: GUIDELINES FOR IDENTIFYING UNDERACHIEVEMENT FACTORS

1. Based on past performance and standardized test scores, does the student have the ability to achieve the grades expected by parents and/or teachers?

2. Is the underachievement spread across all courses? If not, in what subject area(s) is the student not achieving? Has he/she had trouble with this (these) subject(s) in the past? Is he/she at a disadvantage because of inadequate background knowledge for the course(s)?

3. In general, has the student had positive or negative experiences in academic achievement? Does he/she approach learning tasks with confidence, or apprehension? To what does he/she attribute past successes/failures?

4. Does the student know and use effective study/learning strategies? What is his/her learning style?

5. Is the work too difficult for this student? Should he/she be in a different track or taking other courses?

6. Does the student's family provide adequate support/encouragement for his/her academic achievement? Does he live in a community where achievement is valued?

7. Is the student's peer group supportive of academic achievement? To what extent do social or extracurricular activities compete with academic achievement?

8. Does the school provide a supportive atmosphere? Do teachers like this student? Does the student see the school as a supportive, or as a hostile, environment?

9. What is the student's overall disposition? (happy, sad, positive, negative, energetic, lethargic, etc.)

10. Is the student exhibiting adolescent rebellion? Is he/she engaged in a power struggle with parents? Is he/she locked into a developmental stage?

11. Does the student personally value achievement? Does he/she have a clear career goal?

12. Is the student willing to take risks? Does he/she fear failure?

APPENDIX B: STUDENT ACHIEVEMENT ASSETS AND BARRIERS

For each of the following, rate the extent to which the factor is a barrier or an asset to the student's academic achievement. Use the "Comments" areas to note specific strengths or problems.

	Barrier				Asset
Academic Ability	1	2	3	4	5
Comments:					
Specific Prerequisite Knowledge	1	2	3	4	5
Comments:					
Past Learning Experiences	1	2	3	4	5
Comments:					
Study Skills/Learning Style	1	2	3	4	5
Comments:					
Task/Course Difficulty	1	2	3	4	5
Comments:					
Family/Community Support	1	2	3	4	5
Comments:					
Peer Support	1	2	3	4	5
Comments:					

	Barrier				Asset
School Support Comments:	1	2	3	4	5
General Mood/Disposition Comments:	1	2	3	4	5
Psychological Development Comments:	1	2	3	4	5
Values/Career Goals Comments:	1	2	3	4	5
Risk Taking Propensity Comments:	1	2	3	4	5

APPENDIX C: POTENTIAL STRATEGIES FOR OVERCOMING ACHIEVEMENT BARRIERS

Barrier	Strategies

APPENDIX D: ACTION PLAN FOR OVERCOMING ACHIEVEMENT BARRIER

Barrier:

Goal:

Action Step	Person Responsible	Date to be Completed	Evidence	Reward
1.				
2.				
3.				
4.				
5.				

APPENDIX E: TIMELINE FOR MONITORING PROGRESS

Action Step	Projected Completion Date	Actual Completion Date	Reward Given (X)

APPENDIX F: REPORT CARD ANALYSIS
PART 1--MY BEST GRADES: WHY I GOT THEM

For each subject check the *five* reasons that best describe why you were able to get a good grade that semester.

Reason	Subject:	Subject:	Subject:	Subject:	Subject:
I'm good at this subject					
I worked hard					
I kept up on assignments					
I seldom missed class					
I studied for tests					
The teacher was good					
I liked the teacher					
The teacher liked me					
Subject was interesting					
Assignments were easy					
The tests were easy					
Had good luck on tests					

APPENDIX F: REPORT CARD ANALYSIS
PART 2--MY WORST GRADES: WHY I GOT THEM

For each subject check the *five* reasons that best describe why you did not get a good grade that semester.

Reason	Subject:	Subject:	Subject:	Subject:	Subject:
I'm not good at this subject					
I didn't work very hard					
I didn't do the assignments					
I often missed class					
I didn't study for tests					
The teacher was no good					
I didn't like the teacher					
The teacher didn't like me					
Subject was boring					
Assignments were too hard					
The tests were too hard					
Had bad luck on tests					
Didn't have time to study					
Missed school because of health or family problems					

APPENDIX G

(For a free copy of this digest, contact ERIC/CAPS.)

HIGHLIGHTS ...An ERIC/CAPS Fact Sheet

Counseling for Study Skills

Rationale

Counseling and guidance professionals are increasingly serving the learning and developmental needs of all students, rather than the therapeutic or remedial needs of a few. In the area of learning, study skills stand at the top of the list along with reading, writing, mathematics and reasoning. In the area of adolescent development, a number of tasks relate directly to academic achievement; for example, studying effectively, producing in work situations under adult performance standards, and establishing a worker identity. Counselors' expertise in these areas thus makes specialized study skills instruction a logical part of the counseling role at all educational levels.

Research

Research and practice have increased our knowledge of how students learn and do not learn. Because problems in academic performance have been found to relate to study skills deficits and to emotional and personal problems, the complex needs of the student with academic difficulties are best served by an interactive learning system consisting of primary strategies (study skills) and support strategies (counseling). Successful study skills programs incorporate this dual approach by including:

- Study skills instruction combined with counseling.
- Group rather than individual counseling.
- High levels of warmth, empathy, and genuineness.
- Skills instruction related to content material.
- Structured rather than unstructured formats.
- Longer programs (ten hours or more).
- Voluntary participation (at the college level).

Assessment

To design a program capable of meeting these complex needs requires information about students' knowledge and use of general and specific study methods, and about personality characteristics which affect learning. A widely used measure of study skills is the Survey of Study Habits and Attitudes (SSHA, by Brown & Holtzman), a 100-item inventory with four scales measuring habits and attitudes: work methods (use of effective study procedures); delay avoidance (promptness in completing assignments and ability to resist distractions); teacher approval (students' feelings and opinions about teachers); and educational acceptance (students' approval of educational objectives, practices and requirements). Another measure is the Student Attitudes Inventory (SAI, by Entwistle), which has 47 true-false items with four scales: motivation, 14 questions; study methods, 14 questions; examination technique, nine questions; and lack of distractions, ten questions.

Measures of learning style provide information for adapting instruction to personal style. Although several instruments are called the Learning Style Inventory (LSI), each refers to a slightly different view of the concept. LSI by Canfield and Lafferty is a self-report instrument for use with junior high school through adult levels based on a rank ordering of choices for each of 30 questions. Administration time is approximately 15 minutes. It can be used to develop instructional materials for a whole class or for individual students. Its emphasis on attitudinal and affective dimensions makes it a useful tool for counseling.

LSI by Dunn, Dunn and Price is a self-report instrument for use with grades 3-12 based on a rank ordering for each of 104 items. Approximate administration time is 30 minutes. An accompanying manual suggests prescriptions to complement selected styles to facilitate academic achievement.

LSI by Kolb is a self-report instrument for young adults based on a rank ordering of four possible words in each of nine different sets. Approximate administration time is 5-10 minutes. Emphasis is on awareness of personal learning style and available alternative modes.

Another measure of personality is the Myers-Briggs Type Indicator (MBTI) which can be used as an initial screening device. This measure discriminates between those who tend to improve their academic performance with traditional study skills instruction (judgers) and those who are influenced by the amount of course structure provided (perceivers).

Since study skills deficits are often accompanied by poor test performance, a tool for identifying students whose performance is related to personality rather than knowledge is useful. The Achievement Anxiety Test (AAT, by Alpert & Haber) consists of two parts, the facilitating and debilitating anxiety scales, and can be used for this purpose.

Instructional Content

Specific instruction can be developed at all educational levels from basic skill building themes. They include: (1) locating information — using tables of contents, indexes, reader's guides, dictionaries, thesauri, encyclopedias, almanacs, libraries, catalogues, and computerized information retrieval; (2) selecting information — determining main ideas and supporting detail; (3) organizing information — summarizing, notetaking, determining organizational patterns, listening; (4) understanding graphic aids; (5) following simple and complex oral and written directions; (6) developing reading flexibility; (7) remembering information — studying for examinations; (8) using time wisely; and (9) using effective writing skills.

Reading. Underlining, outlining and highlighting are all standard methods for focusing attention and increasing understanding of written texts. The SQ3R and REAP methods require the additional step of processing information in a tangible way and are also widely used.

The SQ3R technique for reading and studying textbooks involves five steps: (1) Survey — glance at chapter headings, read summaries, determine organization; (2) Question — formulate questions about each section to direct further reading; (3) Read — while reading, actively search for answers to formulated questions; (4) Recite — answer questions without reference to the text; and (5) Review — list major

Counseling and Personnel Services Clearinghouse
2108 School of Education, The University of Michigan, Ann Arbor, Michigan 48109-1259 (313)764-9492 1983 ERIC®

subpoints under each heading. Notecards, notebooks and/or tape recorders can be useful adjuncts to this approach.

The REAP reading and study method has four basic steps: Read to discover the message; Encode the message in one's own words; Annotate by writing the message in notes; Ponder the message by processing it through thinking and discussion. Central to the REAP procedure is the process of writing an annotation designed to achieve certain learning objectives. Seven annotation formats have been developed for use with different types of text (summary, thesis, question, critical, heuristic, intention, and motivation).

Notetaking. While fewer systematic hints on how to keep notes have been devised, the 5R's from the Cornell Study Center incorporates the basic processes of effective reading: Record — pick out main ideas; Reduce — summarize, note key terms; Recite — repeat key ideas to oneself; Reflect — think about content; Review — recall and commit to memory.

Test Taking. Instruction in this area involves the following: (1) test preparation — frequent and planful study, adequate rest and diet, blocking out distractions; (2) hints for taking objective and subjective examinations; (3) test wiseness — following instructions, scanning, pacing, reviewing; (4) learning from examinations; (5) managing test anxiety — replacing negative self-statements with positive ones, breathing techniques, progressive relaxation, and desensitization.

Time Management. Common components of time management instruction include: (1) record keeping procedures — daily schedules or diaries to identify habits; (2) schedule planning based on the identified habits and incorporating fixed events; (3) life support activities; (4) leisure time; (5) study time blocked out to allow a commitment for each course; (6) realistic goals for each study session; (7) study breaks; (8) coordination with individual energy periods; and (9) planned use of short time intervals.

Techniques

Behavior modification techniques, which teach people to control their own behavior and change undesirable habits, are readily adapted to individual students and are often applied to infrequent and ineffective studying.

Self-Observation or Self-Monitoring. The learner attempts to observe himself/herself objectively by charting, measuring or counting study behaviors. Equipment can be a simple paper and pencil record or sophisticated computer controlled monitoring. The data serve as a baseline for evaluating change.

Stimulus Control. This technique involves changing the environment. Like all behavior, studying is under some kind of stimulus control, and changing the stimulus will change the behavior. Finding a new, less distracting place to study is an example of this type of environmental change. The knowledge gained from self-observation techniques can help in understanding and changing significant environmental stimuli.

Behavioral Contracts. The learner contracts and administers rewards and punishments based on whether study has been effective or not. The aim of this technique is to increase pre-selected study behavior and reduce undesirable alternatives.

Format

Whether study skills instruction is incorporated into the total curriculum (which is often the case at the middle school and secondary levels) or is a separate course or workshop, a number of formats can be employed:

A combined lecture/peer discussion/practice format focusing on knowledge and use of study skills and emotions allows students to become practiced in active participation in the learning situation, and to overcome anxieties related to academic problems.

Peer tutoring/counseling is an efficient means of providing study skills training to large numbers of students. A successful peer program will require professional supervision, a comprehensive selection process, a good library of study skills books and materials, coordination with counselors and academic advisors or teachers, individual follow-up, and support for the paraprofessional staff.

Programmed and computerized study skills instructions, written instructions and handouts, and audio- and videotapes are also means to reach large numbers of students with the use of fewer personnel. The role of the counselor/instructor in these systems requires assessment, introduction and follow-up, identification of additional resources, and identification and exploration of related personal problems.

The SIP Model

The Study Improvement Program (SIP) offered to second semester college freshmen at the University of Rochester, New York, incorporates the principles suggested here (Malett, 1983, EJ 279 214). In this model, ten paraprofessionals (five men, five women) are selected on the basis of recommendations from staff, faculty, and student leaders; screening interviews, minimum grade point average (2.7, where A = 4.0); and completion of one natural science course.

Training consists of a three-hour seminar once a week (including homework) in basic counseling skills, study skills techniques, and applications to study behaviors of self-control techniques. The seminar format consists of didactic presentations, modeling, practice, and videotaped feedback, with the goal of teaching attending, paraphrasing, questioning, reflection of feelings, interviewing, and related small group discussion techniques. Seminar instructors also model the teaching of study skills and self-control techniques by having the prospective leaders function as a simulated SIP group.

Male-female pairs of SIP leaders conduct the 11 half-hour group sessions comprising SIP. Seven sessions are technique-oriented, directed primarily to teaching behavioral self-control as a study technique, and three sessions are semi-structured discussions of personal factors affecting academic performance. In addition, each group member meets individually with a group leader twice during the semester for a one-hour counseling and problem solving session. The format for individual sessions consists of a review of the previous week's session, formal instruction, practice and homework assignments.

The total 11 sessions are as follows: (1) introduction/self-control techniques; (2) time management; (3) textbook reading efficiency skills; (4) discussion of the importance of grades; (5) stimulus control; (6) test taking and anxiety management; (7) discussion of academic and non-academic pressures; (8) lecture notetaking; (9) discussion of values; (10) writing papers; and (11) problem solving.

Resource Documents

Dunn, R., et al. Learning style researchers define differences differently. *Educational Leadership,* 1981, *38*(5), 372-375. (EJ 242 309)

Eanet, M. G. *An investigation of the REAP reading/study procedure: Its rationale and efficacy.* Paper presented at the 27th Annual Meeting of the National Reading Conference, New Orleans, December 1977. (ED 149 323)

Maher, M. F., & Thompson, M. K. The developmental guidance workshop: Outreach in action. *The School Counselor,* 1980, *28*(1), 39-49. (EJ 233 892)

Malett, S. D. Description and subjective evaluation of an objectively successful study improvement program. *The Personnel and Guidance Journal,* 1983, *61*(6), 341-345. (EJ 279 214)

Rabyak, J. E. A revised study skills model: Do some of them practice what we teach? *The Personnel and Guidance Journal,* 1977, *56*(3), 171-175. (EJ 169 341)

Mary Frenza
Assistant Director for Information Processing, ERIC/CAPS

Tips for Improving Your Listening Comprehension

1. **Be Prepared.** Before going to class, read your textbook assignment and review your notes from the previous class meeting to prepare yourself to get the most out of the lecture.

2. **Concentrate on the Lecture.** Don't allow yourself to be distracted by your professor's mannerisms, voice quality, or delivery technique. Concentrate on the material, not the way it's conveyed.

3. **Listen With an Open Mind.** Maintain a questioning, but open-minded, attitude as you listen. You need not accept everything that is said, but don't let your views cause you to reject a new idea before it is fully developed. Write down your questions and points of disagreement so you won't forget them—and continue to listen to the lecture.

4. **Keep Physically Alert.** Assume a comfortably alert posture in class. Minimize visual or hearing difficulties by sitting up front instead of at the back of the classroom where it is more difficult to pay close attention.

5. **Keep Mentally Alert.** Avoid doodling in class. Concentrate on the lecture, not on the windows, your watch, or the person sitting next to you. Keep attentive in class by predicting likely test questions, comparing lecture and textbook content, entering into class discussion, etc.

6. **Use the Listen-Think-Write Process.** Your professors will be speaking at about 100 words per minute, while you can be thinking at about 400 words per minute—a differential ratio of 1:4! Use this extra time for thinking about what is being said. Train yourself to listen attentively to the material being presented, to evaluate critically

its importance and the evidence to support it. Then select what is appropriate for your notes and record this in your own words.

Tips for Improving Your Recording Skills

1. **Use Outlines**. Whenever possible, take your notes in outline form, using a system of enumeration and indentation to distinguish major and minor points.
2. **Be Neat**. Write legibly and on only one side of the page. This will help you when you review and expand your notes later.
3. **Be Orderly**. Date and identify each set of notes and keep the notes from different courses separated.
4. **Note Study Aids**. Copy diagrams, drawings, and other illustrations that your professor puts on the chalkboard.
5. **Note Examples**. When appropriate, record your professor's examples since they often clarify otherwise abstract ideas.
6. **Note Specific Data**. Make certain that you record correctly all names, dates, places, formulas, equations, rules, etc.
7. **Watch for Emphasis**. Keep alert to points that your professor emphasizes by means of repetition, writing on chalkboard, extended comment, etc.
8. **Listen for Oral Hints**. Keep alert for points your professor emphasizes by means of oral hints. Listen for enumerations such as "the following five steps" or "the four major causes" and for summations such as "consequently" or "therefore." If your professor says, "You'll see this later," or "This is important," follow up such

clues by putting an asterisk or other appropriate symbol in the page margin beside your notes.

9. **Emphasize Notes**. Use symbols, such as an asterisk, a star, or underlining, to indicate points that your professor emphasized.

10. **Clarify Notes**. Draw a circle around reading and other assignments that are mixed in with your lecture notes. Similarly, circle book titles and other references mentioned by your professor.

11. **Note Your Ideas**. Separate your own thoughts from those of your professor. Writing down your own ideas, examples, and questions is an excellent way to keep alert during a lecture. For obvious reasons, however, you should bracket or otherwise label these as your own.

12. **Allow Room to Expand**. Leave enough blank space to permit clarifying and expanding your notes later on. Ask a fellow student or your professor to help you fill in the gaps if you think you missed one or more important points.

13. **Listen Attentively**. Don't be a clock watcher. Your professors cannot always pace themselves accurately, and may cover half of the lecture content in the last ten minutes of class. Pay as close attention to the end of the lecture as you did to the beginning.

14. **Expand Notes**. Review and clarify your notes as soon as possible after class, but do not waste time on recopying them. Use your page margins to fill in abbreviations, add omitted points, correct errors, etc.

Note. Excerpted from *A guide to college survival, revised edition* (pp. 50-52) by W. F. Brown and W. H. Holtzman, 1987, Iowa City, IA: The American College Testing Program.

APPENDIX I: TIPS FOR TAKING MULTIPLE-CHOICE TESTS

Multiple-choice Questions

Check the directions to see if the questions call for more than one answer.

Answer each question in your head before you look at the possible answers. If you can come up with the answer before you look at the choices, you eliminate the possibility of being confused by those choices.

Mark questions you can't answer immediately and come back to them if you have time.

If you have no clue as to what the answer is, and if incorrect answers are not deducted from your score, use the following guidelines to guess:

1. If two answers are similar, except for one or two words, choose one of these answers.
2. If two answers have similar sounding or looking words (intermediate—intermittent), choose one of these answers.
3. If the answer calls for a sentence completion, eliminate the answers that would not form grammatically correct sentences.
4. If two quantities are almost the same, choose one.
5. If answers cover a wide range (4.5, 66.7, 88.7, 90.1, 500.11), choose one in the middle.
6. If there is no penalty for guessing and none of the above techniques work, close your eyes and go for it.

Note. Excerpted from *Becoming a master student* (p. 153) by D. B. Ellis, 1985, Rapid City, SD: College Survival, Inc.